DICK DEKKER

Hunting tactics of Peregrines and other falcons

ISBN978-0-88839-683-9
Copyright © 2009 Dick Dekker

Cataloging in Publication Data

Dekker, Dick, 1933–
    Hunting tactics of peregrines and other falcons / Dick Dekker.

    Originally presented as the author's thesis (doctoral-Wageningen University).
    Includes bibliographical references.
    Also available in electronic format.
    Includes some text in Dutch.

    ISBN 978-0-88839-683-9

    1. Peregrine falcon—Canada, Western. 2. Predation (Biology).
3. Falco—Canada, Western. I. Title.

QL696.F34D352 2009   598.9'6   C2009-900606-5

All rights reserved. No part of this publication may be reproduced, stored in a retrieval system or transmitted, in any form or by any means, electronic, mechanical, photocopying, recording, or otherwise, without the prior written permission of Hancock House Publishers.

Paintings, graphic design and editing (English and Dutch) by Dick Dekker

Printed in the USA

We acknowledge the financial support of the Government of Canada through the Book Publishing Industry Development Program (BPIDP) for our publishing activities.

Published simultaneously in Canada and the United States by

HANCOCK HOUSE PUBLISHERS LTD.
19313 Zero Avenue, Surrey, B.C. Canada V3S 9R9
(604) 538-1114 Fax (604) 538-2262

HANCOCK HOUSE PUBLISHERS
#104-4550 Birch Bay-Lynden Rd, Blaine, WA U.S.A. 98230-9436
(800) 938-1114 Fax (800) 983-2262

Website: **www.hancockhouse.com**
Email: **sales@hancockhouse.com**

Hunting tactics of Peregrines and other falcons

Proefschrift

Ter verkrijging van de graad van doctor
op gezag van de Rector Magnificus
van Wageningen Universiteit
Prof. Dr. M.J. Kropff
in het openbaar te verdedigen
op woensdag 18 februari 2009
des namiddags te vier uur in de Aula.

Dekker, D.
Hunting tactics of Peregrines and other falcons.

Ph.D. thesis, Wageningen University,
Wageningen, The Netherlands
with a summary in Dutch.

Promotoren

Prof. Dr. R. C. Ydenberg
Hoogleraar Faunabeheer,
Wageningen Universiteit

Prof. Dr. H.H.T. Prins
Hoogleraar Resource Ecology,
Wageningen Universiteit

Promotiecommissie

Dr. W. Cresswell
University of St. Andrews, Scotland

Prof. Dr. R.W. Goldbach
Wageningen Universiteit

Dr. H. Hakkarainen
University of Turku, Finland

Dr. B. Nolet
Nederlands Instituut voor Ecology (KNAW), Maarssen

Dit onderzoek is uitgevoerd
binnen de onderzoekschool C.T. de Wit Graduate School
for Production Ecology and Resource Conservation

TABLE OF CONTENTS

Abstract 6

Voorwoord 9

Samenvatting 13

Foreword 16

Chapter 1. Introduction, study areas, and methods 19

Chapter 2. Peregrine migrations in Alberta 30

Chapter 3. Hunting strategies of migrating Peregrines 41

Chapter 4. Duck hunting by migrating and wintering Peregrines 60

Chapter 5. Foraging habits of wintering Peregrines 66

Chapter 6. Marine Peregrines hunting seabirds 72

Chapter 7. Over-ocean flocking by wintering Dunlins 77

Chapter 8. Peregrine predation on Dunlins in relation to ocean tides 85

Chapter 9. Peregrine prey selection and eagle interference 92

Chapter 10. Cooperative hunting by parent Peregrines and their fledglings.. 101

Chapter 11. A comparison of Peregrines and Merlins hunting small
    shorebirds and passerines 115

Chapter 12. Nest site competition between Peregrines and Prairie Falcons ... 123

Chapter 13. Gyrfalcons and Prairie Falcons hunting city pigeons 135

Chapter 14. Gyrfalcons and Bald Eagles hunting wintering Mallards 144

Chapter 15. Herfsttrek en jachtgewoontes van de Slechtvalk in Friesland ... 150

Chapter 16. Discussion and synthesis 163

Acknowledgements 177

Literature Cited 179

Curriculum Vitae 192

ABSTRACT

This dissertation describes the foraging habits and capture rates of four species of bird-hunting falcons; Peregrine (*Falco peregrinus*), Merlin (*Falco columbarius*), Gyrfalcon (*Falco rusticolus*), and Prairie Falcon (*Falco mexicanus*). Eight of the nine study areas were situated in western Canada in widely different habitats, and the observation periods intermittently included all seasons over 44 years, 1965–2008. The various chapters report and compare the location-specific hunting methods and choice of prey of these falcons in the following scenarios: (1) Migrating Peregrines hunting waterbirds at Beaverhills Lake, a large wetland in central Alberta; (2) Migrating Peregrines and Merlins capturing small shorebirds and passerines at Beaverhills Lake; (3) Breeding Peregrines that launched their hunts from the high chimneys of an industrial powerplant by a large Alberta lake and selectively took gulls; (4) Marine Peregrines nesting on Pacific island cliffs and preying on seabirds; (5) Peregrines and other raptors hunting wintering Dunlins (*Calidris alpina*) and ducks at Boundary Bay on the Pacific coast of British Columbia; (6) Peregrines specializing on teal and American Wigeon (*Anas americana*) wintering on coastal farmlands; (7) Territorial pairs of Peregrines and Prairie Falcons competing for prey and nest sites on a sympatric breeding range along an Alberta river; (8) Prairie Falcons and Gyrfalcons wintering in the city of Edmonton and capturing Rock Doves (*Columba livia*); (9) Gyrfalcons hunting Mallards (*Anas platyrhynchos*) wintering on Alberta farmlands; (10) Klepto-parasitic interference from eagles and buteo hawks with hunting Peregrines and Gyrfalcons, and intra- and interspecific prey theft between falcons.

The largest set of data pertains to the Peregrine, which was studied in all seasons and habitats except during winter in central Alberta. At Beaverhills Lake, Peregrines on spring migration attacked avian prey on 674 occasions and captured 52, a success rate of 7.7%. Success rates for adult falcons and spring immatures were respectively 9.8% and 7.1%, not significantly different. Fall immatures had a success rate of 2.4%, but the sample was small. Waterfowl and shorebirds made up 94% of prey taken. Stealth approaches and long-range surprise attacks were the major strategies in 70% of hunts. Hunting Peregrines often soared at great altitudes and stooped low to attack flying prey or to flush sitting prey. The effectiveness of surprise strategies at the lake was facilitated by shoreline vegetation. In addition, Peregrines used a number of other, more unusual, hunting methods. A few prey were taken on the ground or in shallow water, all others were seized in the air and borne down. The majority of birds caught failed to use escape tactics routinely employed by their kind. Lone prey individuals were more often killed compared to individuals in flocks.

Peregrines wintering on the Pacific coast attacked Dunlins 652 times with 94 captures, representing an overall success rate of 14.4%. Adult falcons achieved 26.8% success in 164 hunts, significantly higher than the 9.0% in 399 hunts by first-year Peregrines. The hunting strategies of these wintering falcons differed from the tactics deployed at Beaverhills Lake. The open coastal mudflats did not facilitate a stealth approach. Only 35% of Peregrine hunts at Boundary Bay were surprise attacks and 62% were open attacks on flocks. Hunts over the mudflats or ocean had a success rate of 11%, as compared to 44% over the shore zone. Some very long and persistent pursuits of sandpipers were made by immature Peregrines and Merlins.

During high winter tides that inundated all mudflat habitats, the Dunlins departed and stayed well offshore for periods of 1.5 to 6.5 hours. Termed over-ocean flocking, this flight behaviour is believed to be an anti-predator strategy. After the tide began to recede the Dunlins landed in the shore zone where they were at their most vulnerable to surprise attacks by Peregrines. Captures recorded per hour of observation peaked at 0.25/hour in the two hours following the crest of the high tide, four times higher than two hours prior to high tide. The hypothesis is advanced that nearness to vegetation increases the Dunlin's predation risk, which was strongly supported by the data.

The hypothesis that the hunting success rate of breeding Peregrines is greater than that of migrating or wintering falcons was supported by data collected at an Alberta nest site. The adult pair had an overall rate of 30.3% in 386 hunts. Over the course of the 7-year study the rate increased from 21.9% in the first year to 39.1% in the seventh year. The majority (77%) of attacks were initiated from high soaring flight, and the falcons used the exhaust of the plant's smokestack to gain height for soaring. Of 117 prey captured 62 (53%) were Franklin's Gulls (*Larus pipixcan*), and 85% of a total of 81 gulls known to have been killed by these falcons were juveniles.

Peregrines breeding on an island off the northwest Pacific coast hunted seabirds such as murrelets by direct aerial pursuit or by striking them on the water. If the swimming bird was hit and crippled by the attack, the falcon turned back to pick it up from the surface of the water. The success rate of these falcons was 22% in 73 observed hunts.

In hunting small passerines and small shorebirds, the Peregrine's success rate was 8.2% in 647 hunts as compared to 12.4% in 354 hunts by Merlins. The Merlin was significantly more successful than the Peregrine in capturing small passerines (12.2% vs. 3.8%) but not in hunting small shorebirds (12.6% vs. 8.8%).

The majority of ducks (82%) taken by Peregrines were, in order of frequency, American Wigeon, Northern Pintail (*Anas acuta*), Northern Shoveler (*Anas clypeata*), and teal (*spp.*). Mallard hens were rarely captured but no drakes were seen killed. Mallards of both sexes were preyed upon by Gyrfalcons wintering in Alberta. Their success rate in 70 hunts was 22.8%. Gyrfalcons wintering in Edmonton made 141 attacks on Rock Doves with a success rate of 10.6%, significantly different from the 26.0% success of a Prairie Falcon in 104 attacks on city pigeons.

Klepto-parasitic interference was commonplace between falcons and larger raptors. In British Columbia, Peregrines and Gyrfalcons lost captured ducks to Bald Eagles (*Haliaeetus leucocephalus*) and Golden Eagles (*Aquila chrysaetos*). In Alberta and the Netherlands, Peregrines were robbed by buteo hawks. As well, interspecific and intraspecific prey theft was frequent between falcons, and females routinely robbed conspecific males.

In addition, this thesis reports on the population dynamics and nest site competition between Peregrines and Prairie Falcons on a sympatric breeding range in Alberta over a span of 48 years. The Peregrine became extirpated during the 1960s. Large-scale efforts to reintroduce the species in the 1990s seemed initially successful, resulting in seven new breeding pairs, but they dwindled to one, while the Prairie Falcons continued to do well.

The first chapter of the thesis details 15 years of Peregrine migrations at a large lake in central Alberta, and the last chapter analyses the fall passage and hunting habits of Peregrines along the Wadden Sea coast of the Netherlands. There, the specific research question was whether or not Dunlins wintering on the Dutch coast would engage in over-ocean flocking manoeuvres during high tides that inundate all mudflat habitats. The phenomenon proved to be very rare and the probable reasons are discussed in detail.

VOORWOORD

"Dekker, loop niet zo stom naar de lucht te kijken!" schreeuwde de sergeant toen ik in mijn peleton dienstplichtigen langs de grote kerk van Amersfoort marcheerde. Het gebeurde meer dan vijftig jaar geleden, maar ik herinner me het voorval als de dag van gisteren. Bij de imposante Dom vloog een dichte klucht duiven en erboven cirkelde een Slechtvalk, klaar voor de stoot. Hoe het afliep ben ik nooit te weten gekomen.

Alhoewel er in die tijd nog geen Slechtvalken in Nederland broedden, vormden de vogelrijke Lage Landen van oudsher een geschikt overwinteringsgebied voor noordelijke trekvalken. Die zaten graag op hoge uitkijkposten zoals kerktorens. Als jonge vogelaar vond ik de binnenstad echter allerminst ideaal voor het observeren van een fascinerende "roofpiet" als de Slechtvalk. Ik fietste liever door de polders nabij mijn woonplaats Haarlem. Daar nam ik tenslotte voor het eerst waar hoe een Slechtvalk zijn prooi ving. Tot mijn verbazing verliep dat heel anders dan de typische beschrijvingen in de vogelboeken, waarin algemeen werd beweerd dat deze "edele valk" zijn prooi in snelle vlucht achterhaalde en hem dan, in een steile stoot, een klap met de klauwen gaf zodat het slachtoffer dood of gewond omlaag tuimelde.

Op een mistige wintermiddag onthulde mijn verrekijker echter een heel ander scenario. Een volwassen Slechtvalk die al een tijdje in een hoge mast gezeten had vloog plotseling schuin naar beneden en vervolgens laag over de grond om een zittende Smient te grijpen, gewoon op de natte wei. Tot mijn verdere verbazing zag ik later dat een Slechtvalk een Sperwer beroofde van een pas gevangen Koperwiek. Blijkbaar is deze valk, die bekend staat als het snelste dier ter wereld, liever lui dan moe. Hij pakt dan gewoon wat het makkelijkste is. Deze ontdekking wakkerde mijn jeugdige passie aan en verscherpte mijn plan om van nu af aan een gedetailleerd logboek bij te houden van mijn observaties. Ik wilde niet alleen precies weten hoe de valk zijn buit ving, maar ook hoe prooivogels het klaarspeelden zo'n verraderlijke vijand te ontwijken.

Mensen die een voorliefde aan de dag leggen voor roofdieren lopen het risico verdacht te worden van een sadistische inslag. Maar zo ervaar ik dat helemaal niet. Verre van mij verheugen over het lot van de slachtoffers, komt mijn betrokkenheid voort uit een gesublimeerd jachtinstinct. In plaats van zelf te jagen, doet de valk dat als het ware voor mij. Ik leef mee *vicariously*. De extase van de jacht, het zich meten in snelheid met een ander, zit er bij ons van kinds af aan in en verklaart de populariteit van voetbal. De bal fungeert als prooi. De adrenaline verhoging dat het bloedloze spel veroorzaakt in zelfs de meest evenwichtige en moderne stadsmens, komt tot uiting bij elke belangrijke match. Persoonlijk kan ik mij over voetbal niet al te druk maken, maar als ik de verrekijker richt op een valk in felle achtervolging van een duif of eend, hoog in de lucht, heb ik moeite niet hardop te juichen. Ik kan het dan ook niet vaak genoeg zien.

Als een liefhebber van deze toeschouwerssport, als ik het zo mag noemen, verkeer ik in goed gezelschap. De valkenjacht met afgerichte vogels was vroeger "the sport of kings." Tot vermaak van vorstenhoven stond het vederspel meer dan duizend jaar lang in hoog aanzien, totdat de uitvinding van het jachtgeweer de valk van zijn voetstuk stootte.

Het houden van roofvogels beleeft tegenwoordig een nieuwe opbloei, maar er zijn maar weinig liefhebbers in staat, of in de tijdrovende gelegenheid, de valk eer aan te doen in de klassieke hoge vlucht. Destijds, in de vijftiger jaren, bevriende ik een van de Nederlandse grootmeesters, Henk Dijkstra, die in 2006 overleed. Wij deelden een levenslange passie voor de Slechtvalk, alhoewel we van meet af aan verschillende wegen bewandeld hebben. Henk leende me ook Engelse boeken over de valkerij, die toen ver vooruit lagen op wat er in Nederland in de handel was. Vergeleken met de overvloed aan materiaal dat de jeugd van vandaag tot zijn beschikking heeft, werd er in die vroege na-oorlogse jaren vrijwel niets gepubliceerd over de natuur, laat staan over roofvogels, die toen een slechte naam hadden.

Legaal of illegaal werd alles wat een kromme snavel of klauwen had bestreden als schadelijk wild en ongewenste competitie voor de jager. De verdrukte roofvogelstand zonk naar een dieptepunt in de natuurramp die plaatsvond in de zestiger jaren, veroorzaakt door de giftige residu's van land- en tuinbouwchemicaliën. De desastreuze gevolgen van die periode zijn welbekend. Als broedvogel verdween de Slechtvalk uit bijna geheel Europa en Noord America (Hickey 1969). Het verrijzen van de soort werd vervolgens een tweede fenomeen van wereldformaat. De Slechtvalk is nu weer thuis in alle vroegere broedgebieden (Cade and Burnham 2003), en mede door beschermingsmaatregelen, breidt zijn areaal zich gestadig uit. Het aantal nesten in Nederland, waar de soort voorheen niet of zeer zelden broedde, is nu opgelopen tot over de dertig (Van Geneijgen 2006).

In de nasleep van de milieucrisis uit de tweede helft van de vorige eeuw heeft de Slechtvalk een spectaculaire metamorfose doorgemaakt. Van obscuriteit – als een vogel waar heel weinig mensen ooit van gehoord hadden – staat hij nu algemeen bekend als een symbool van het herstel van de natuur. De terugkeer van de valk is de kroon op het werk van velen die zich ingespand hebben om het pesticidengevaar te weren. Roofvogels en hun predatie zijn ook in een nieuw licht komen te staan in de ornithologische gemeenschap. Voorheen, vanwege hun zeldzaamheid, werd er nauwelijks aandacht besteed aan valken in het Nederlandse natuurgebied bij uitstek, de waddenkust. Omstreek 1985 herinner ik mij een ontmoeting met een veldbioloog van de Universiteit van Groningen, ergens in een groene polder bij het Lauwersmeer. Zijn telescoop stond gericht op ganzen. "Slechtvalken?" zo reageerde hij in antwoord op mijn vraag, "Daar kijken we niet naar."

Deze afwijzende of neutrale houding betreffende roofvogels veranderde grondig toen men zich begon af te vragen wat de directe en indirecte gevolgen waren van de spectaculaire terugkeer van de Slechtvalk. Studies naar de invloed van de toegenomen predatiedruk op het gedrag en de morfologie van steltlopers staan nu op de voorgrond (Piersma et al. 2003; Ydenberg et al. 2004).

Een van de eersten die zich met deze vraag bezig hield was Professor Ronald Ydenberg van de Centre for Wildlife Ecology van Simon Fraser University in Vancouver, Canada. Alhoewel ik al in 1994 begonnen was met mijn intensieve observaties aan Slechtvalken op de Canadese westkust, duurde het tot 2000 voor ik het genoegen had met Ron kennis te maken. De ontmoeting vond plaats in zijn kantoor in Delta, British Columbia, op het Pacific Wildlife Research Centre van de Canadian Wildlife Service.

Ron en zijn collega's hadden al jarenlang studie gemaakt van de doortrekkende en lokaal overwinterende strandlopers. De recente terugkeer van de Slechtvalk werd beschouwd als van cruciaal belang, maar gegevens over predatiedruk ontbraken. En dit was juist mijn specialiteit. Onze samenwerking viel in goede aarde en culmineerde in een gezamelijke publicatie die deel uitmaakt van dit proefschrift. Een van de nevenresultaten was ook dat ik in de winters van 2006 en 2007 een kijkje kon gaan nemen in Friesland waar de Slechtvalk nu eindelijk weer zijn naam eer aandoet. Het woord *slecht*, nu van een heel andere gevoelswaarde, betekende vroeger *algemeen*.

De samenwerking met Ron was voor mij ook van financieel belang. Hij nam een deel van de kosten voor zijn rekening als ik naar de kust reisde. Voordien waren mijn zeer bescheiden aanvragen bij overheidsinstanties altijd afgewezen. Van de British Columbia Wildlife Department kreeg ik als antwoord: *"We don't fund people from out of province."* Als ik dan mijn verzoek richtte aan de Alberta Fish and Wildlife Department was de reactie omgekeerd maar even negatief: '*We do not fund out of province research.*" Bij de federale regering was het steevaste refrein: *"No money..."*

Mijn onderzoek heeft nooit afgehangen van anderen. Zolang ik kon doen waar ik zin in had en waarvan ik het belang inzag, had ik daar mijn eigen tijd en geld ruimschoots voor over. Op een paar uitzonderingen na, is mijn reizen en trekken bekostigd uit eigen zak. Dankzij een uiterst spaarzame aard en een even spaarzame vrouw kon ik het financieel net redden. Het belangrijkste was dat Irma me altijd blijmoedig liet gaan, zodat ook onze zoon Richard het normaal vond dat zijn vader vaak de natuur in trok. We hadden ook het grote voordeel dat ik als een freelance grafisch ontwerper en schrijver mijn werk thuis kon doen.

Het resultaat was dat ik niet gebonden was aan beperkingen opgelegd door werkgevers en instituten. Een veldonderzoek kon ik zo lang volhouden als ik wilde, in tegenstelling tot de doorsnee universiteitsstudent die drie of vier jaar krijgt om een project af te ronden. Mijn eerste publicaties, over Slechtvalken bij Beaverhills Lake, zagen pas het licht nadat ik 15 trekseizoenen, zonder uitzondering, in het veld doorgebracht had. Dit verklaart het enorme aantal gegevens, die ver overtroffen wat de bestaande literatuur destijds te bieden had. In elk nieuw studiegebied begon ik met een schone lei, gewoon zonder vooropgezet plan of hypothese, maar wel met open oog. Dit leidde tot originele ontdekkingen van valkengedrag waarover niets of weinig bekend was, of waarvan de bestaande beroepsliteratuur een foutieve of onvolledige voorstelling gaf.

"Helemaal vrij zijn om te onderzoeken waar je nieuwsgierig naar bent is een ongekende luxe," schreef Patrick Jansen, een recent afgestuurde Wageningse bioloog, in zijn dissertatie. Van zo'n vrijheid heb ik met volle teugen genoten en ik hoop dat het resultaat lezenswaard is, dat ik iets heb kunnen bijdragen aan de kennis over de Slechtvalk en andere valken en hun invloed op prooivogels, alsmede hun onderlinge verhouding met elkaar en andere roofvogels.

Edmonton, 21 June 2008.

Plumage differences between North American and European Peregrines are only minor and show individual variation. Unlike this female of British origins, other adult falcons have unmarked white breasts and a much wider malar bar. (Photo: Dick Dekker).

SAMENVATTING

Dit proefschrift beschrijft de jachtwijze en succespercentages van vier soorten valken die op vogels jagen: Slechtvalk (*Falco peregrinus*), Smelleken (*Falco columbarius*), Giervalk (*Falco rusticolus*), en Prairievalk (*Falco mexicanus*). Het onderzoek vond plaats in alle seizoenen en in diverse biotopen in het westen van Canada verdeeld over een tijdperk van 44 jaar, 1965 tot 2008. De 15 hoofdstukken gaan over de manier van jagen en prooikeuze van deze valken in de volgende scenario's: (1) Slechtvalken op doortrek bij Beaverhills Lake, een uitgestrekt moerasmeer in een landbouwstreek in centraal Alberta; (2) Slechtvalken en Smellekens die op strandlopers en kleine zangvogels jagen bij Beaverhills Lake; (3) Een Slechtvalkenpaar dat selectief op meeuwen jaagt en in een nestkast broedt op een electriciteitscentrale gelegen aan een groot Alberta meer; (4) Slechtvalken die ten noordwesten van British Columbia op een oceaaneiland broeden en zeevogels jagen; (5) Slechtvalken die 's winters jacht maken op Bonte Strandlopers (*Calidris alpina*) en eenden in Boundary Bay, een waddengebied gelegen aan de kust van British Columbia; (6) Overwinterende Slechtvalken die zich specialiseren op talingen en Smienten (*Anas americanus*) in een agrarische gebied op Vancouver Island; (7) Slechtvalken en Prairievalken die met elkaar strijden over nestplaatsen op steile rotswanden langs een rivier in centraal Alberta; (8) Een vergelijking tussen de jachtmethodes van Giervalken en een Prairievalk die 's winters stadsduiven (*Columba livia*) vangen in Edmonton, Alberta; (9) De jachtmethodes van Giervalken die in een landbouwgebied bij Edmonton op overwinterende Wilde Eenden (*Anas platyrhynchos*) jagen; (10) Prooiroof door de Amerikaanse zeearend (*Haliaeetus leucocephalus*) ten koste van Slechtvalk en Giervalk, alsmede intra- en interspecifieke prooiroof tussen valken onderling en andere roofvogels.

De langste lijst gegevens over de Slechtvalk werd verzameld in Alberta, van april tot october, en in diverse biotopen. Bij Beaverhills Lake maakten trekkende valken 674 aanvallen op prooivogels waarvan er 52 gevangen werden, een succespercentage van 7,7%. De percentages voor volwassen en onvolwassen valken op voorjaarstrek lagen respectievelijk op 9,8% en 7,1%. Statistisch was het verschil onbelangrijk. Tijdens de herfsttrek was het succespercentage van onvolwassen valken 2,4%, maar hun aantal jachtvluchten was gering. De geslagen prooien bestonden voor 94% uit eenden en strandlopers. De meest gebruikte strategie (70%) in dit gebied was een lage verrassingsaanval. Jagende Slechtvalken, op zoek naar prooi, schroefden vaak op grote hoogte en maakten steile stootduiken om vogels te overrompelen die op de grond en in ondiep water zaten. De doeltreffendheid van verrassingsaanvallen werd bij het meer begunstigd door de oeverbegroeiing van riet en biezen. Een klein aantal prooien werd op de grond of in ondiep water gepakt, alle andere werden in de lucht gegrepen net nadat ze opvlogen. Behalve deze verrassingstactiek hadden de valken nog verschillende andere methodes. Het merendeel van de gevangen strandlopers vertoonde abnormaal ontsnappingsgedrag voor hun soort. Een vogel alleen had meer kans gegrepen te worden dan vogels in groepen.

Slechtvalken die in Boundary Bay overwinterden maakten 652 aanvallen op Bonte Strandlopers en vingen er 94, een percentage van 14,4%. Volwassen valken hadden succes in 26,8% van 164 pogingen, statistisch gezien een belangrijk verschil met eerstejaars valken die een vangstpercentage van 9,0% boekten in 399 aanvallen. De jachtstrategie van deze valken was hoofdzakelijk anders dan bij Beaverhills Lake, hetgeen te maken had met het feit dat het kale kustbiotoop minder geschikt was voor verrassingaanvallen. Slechts 35% van alle jachtvluchten in Boundary Bay begon als een poging de prooi bij verrassing te pakken. Ter vergelijking waren 62% van de aanvallen gericht op groepen strandlopers die op het open wad zaten of over de oceaan vlogen. Vluchtende strandlopers werden soms over lange afstanden en hoog in de lucht achtervolgd door jonge valken, meestal zonder resultaat omdat de prooi tenslotte dekking zocht in ruigte. Over het wad of de oceaan had de Slechvalk een succespercentage van 11%, vergeleken met 44% bij de begroeide kwelderrand.

Tijdens hoog water, als het gehele wad onderstroomde, bleven de strandlopers urenlang boven de oceaan rondvliegen, vaak ver uit de kust. Dit gedrag werd *over-ocean flocking* genoemd en beschouwd als een anti-predator strategie. Als de strandlopers bij vallend getij weer neersteken bleken zij het meest kwetsbaar voor verrassingsaanvallen en beliep het aantal vangsten 0,25 per observatieuur, vier maal zo veel als in de twee uur voor hoogwater. Deze vangstcijfers stonden in correlatie met de hoogte van het getij en bevestigden de hypothese dat de strandlopers dichtbij de oevervegetatie het meeste gevaar liepen overrompeld te worden.

Een zevenjarig onderzoek bij een nestkast wees uit dat het jachtsucces van Slechtvalken in hun broedterritorium groter is dan dat van trekkende of overwinterende valken. Het valkenpaar had een gemiddeld success van 30.3% in 386 jachtvluchten. Over de tijdsduur van de studie nam het succespercentage toe van 21,9% in het eerste jaar tot 39.1% in het zevende broedseizoen. Het merendeel (77%) van de aanvallen op prooi werd ingezet vanuit hoog schroevende zeilvlucht en om snel hoogte te winnen maakten de valken gebruik van de opstijgende uitlaatgassen van de schoorstenen. Iets meer dan de helft (53%) van de 117 gevangen prooien waren meeuwen (*Larus pipixcan*) en hiervan waren er 85% in onvolwassen verenkleed.

Slechtvalken die op een rotseiland in de oceaan broedden in het noordwesten van British Columbia joegen op zeeduikers, zoals kleine alken, die ze bemachtigden door laag over de zwemmende vogel te vliegen en ze in het voorbijgaan een klap met de klauwen te geven. Als de alk geraakt en gewond was, werd hij uit het watervlak gevist. Het jachtsuccess van deze valken was 22% in 73 pogingen.

In de jacht op kleine zangvogels (*emberizidae*) en strandlopers had de Slechtvalk een success van 8.2% in 647 aanvallen vergeleken met 12.4% in 354 aanvallen van het Smelleken. Deze kleinere valk had statistisch gezien meer succes dan de Slechtvalk in de jacht op zangvogeltjes (12,.2% versus 3,8%) maar niet, statistisch bekeken, in de jacht op strandlopers (12,6% versus 8,8%).

Het merendeel van de door Slechtvalken geslagen eenden (82%) bestond uit de volgende naar prooiaantallen gerangschikte soorten: Smient, Pijlstaart (*Anas acuta*), Slobeend (*Anas clypeata*), en taling (*spp.*). Van de Wilde Eend werden een paar vrouwtjes gepakt maar nooit een woerd. Wilde Eenden van beider geslacht vielen wel geregeld als slachtoffer van overwinterende Giervalken, die 22,8% succes boekten in 70 overrompelingsaanvallen en lange achtervolgingen. Giervalken die in de stad overwinterden maakten 141 aanvallen op duiven met 10,6% succes, een statistisch belangrijk verschil met het 26,0% percentage van een Prairievalk, die 104 aanvallen op de duiven deed.

Prooiroof en aggressieve achtervolgingen van prooidragende valken door grotere roofvogels waren routine. Aan de westkust raakten Slechtvalken en Giervalken de door hen gevangen eenden kwijt aan de plaatselijk talrijke Zeearend en soms aan de minder algemene Steenarend (*Aquila chrysaetos*). In Alberta en ook in Friesland werden Slechtvalken lastig gevallen door buizerds. De grotere valken beroofden kleinere soorten en intraspecifiek ging het altijd ten koste van de mannetjes.

Deze dissertatie bevat ook een hoofdstuk over de concurrentiestrijd tussen Slechtvalk en Prairievalk in een gezamelijk broedgebied langs de Red Deer River in Alberta, waar het aantal nestparen van beide soorten aan grote schommelen onderhevig is geweest in de periode van 1960 tot 2008.

Het laatste van de 15 hoofdstukken gaat over doortrekkende Slechtvalken op de Waddenzeekust van Friesland. Hier werd speciaal aandacht besteed aan de mogelijkheid dat de lokaal overwinterende Bonte Strandlopers ook *over-ocean flocking* gedrag zouden vertonen net als in Canada. Dat bleek zeer zelden het geval te zijn en de redenen daarvoor worden besproken.

FOREWORD

"Why this interest in Peregrines?" asked the prominent Dutch ecologist Herbert Prins when I first met him on the great sea wall of Friesland. It was a visceral question.

Here I was back in my country of birth after an absence of nearly half a century, and I was spending day after day on this bleak coast, in November, when the bone-chilling winds hardly ever stopped blowing. There were ducks galore and four species of wild geese passing overhead, but my singular quest was to see the Peregrine.

My answer came without hesitation. "As a boy, growing up in this crowded, human-dominated environment of Holland, birds of prey represented to me all that was truly wild and natural."

Herbert said that he could appreciate and identify with the sentiment. Driven by a similar longing for wild nature, he had sought out the large mammal fauna of Africa. I, by contrast, had turned my compass into the opposite direction, to the north, in search of boreal wilderness, of wolves, eagles, and falcons.

Those who profess to a liking of predators run the risk of being suspected of a sadistic streak. As a student of all creatures of tooth and claw, I have been looked at askance by hunters, who appear unaware of their own paradox that they kill what they love. Personally, I have no quarrel with responsible hunters. In fact, in nature appreciation, I may be closer to the lone wildfowler than to the birdwatching crowd. The way I see it, my quest for falcons derives from a sublimated hunting instinct. In my bloodless kind of pursuit, the birds of prey, as it were, do my hunting for me.

Far from gloating over the fate of their victims, my fascination with falcons centres on the techniques of the hunt. But of equal interest are the reactions of the prey species, how they strive to avoid being captured, whether they hide or flee. How a Peregrine catches a sandpiper is of no more pertinence than the question of how sandpipers cope with a spectacular enemy that may attack at any moment and can do so many times each day.

The loosening of the predator-prey bondage, as a result of the extermination of raptors and large carnivores, has been a phenomenon of comparatively recent times. Nowadays, in Holland and elsewhere, great flocks of ducks and geese travel to and from their wintering grounds without the escort of their nemesis, the eagle. Equally unbalanced and unnatural is the ecology of hoofed mammals when neither wolf nor wildcat is allowed to exact its bloody toll. In Holland, herds of deer live by the grace and at the mercy of mankind under conditions that are far from ideal.

In 1956, daydreaming of places where the ecological balance had not yet been violated to the same extent as in the Netherlands, I travelled to the sub-arctic tundra of Scandinavia. But even here, the large mammal system was a managed one. During the year when I hiked the *Kung's Leden* in Swedish Lapland, the last wolf of the region had been tracked down and shot from a helicopter. The machine's rental was paid for by the high bounty on dead wolves offered by the local reindeer herders.

My summer in Sweden, exploring the woods and lakes of Smaoland, reinforced my tendency to walk alone. Tuning in on the wavelength of predators and prey requires an alert yet relaxed state of mind, which leaves little room for other hobbies such as wildlife photography. Carrying a camera, one easily becomes preoccupied with the subject's merits as a picture. The conflicting demands of photography and fruitful observation became clear to me when I was sitting quietly on a rocky islet in the Baltic Sea. A hoped-for opportunity became a reality when a White-tailed Eagle approached on a course that would take it over a raft of eider ducks within good viewing distance from my place of concealment. But instead of remaining still and feasting my eyes on the majestic bird, I grabbed the camera with the long telephoto lens. The eagle spotted the movement and turned away in a hurry. Another time, I saw a Sparrowhawk grab a Fieldfare, but my clumsy attempt at getting into camera range, caused the hawk to release its wounded victim. Instead of looking through the viewfinder of a camera, I now just watch and listen, jotting down my impressions at the end of the day in bulging notebooks.

In 1959, my quest for a landscape where plant and animal associations were still complete and not greatly disrupted by humans, finally led me to Canada. In 1961, during a wilderness trip in the Yukon, the canoe capsized and my camera equipment ended up on the bottom of a turbulent river. I never replaced the big telephoto lens. It was the least of my losses. The accident could have cost me my life as well. For details of those Canadian adventures, see Dekker (1998, 2001).

At the time of my difficult decision to emigrate to Canada, one of the greatest environmental catastrophes ever – the pesticide menace – had just exploded onto public consciousness. Residues of chemicals used in agriculture and forestry were having a lethal effect on birds of many species, and the kinds that were harmed most were those at the top of the food chain, the predators. Birdwatchers and ornithologists who were around at that time, half a century ago, will recall the panic and worry of that era. I for one felt particularly alarmed. The one bird I valued most had become a martyr of the human rage to meddle with nature. It is now common knowledge that the breeding population of the Peregrine collapsed in much of the human-inhabited world (Hickey 1969). On the brighter side, it is equally well-known that the return of the Peregrine became one of greatest victories for conservation, and thanks are due to all those who worked passionately to resolve the insidious pesticide threat and to bring back this now celebrated falcon from the brink of extinction (Cade and Burnham 2003).

In 1964, when I moved to Edmonton in central Alberta, Peregrine Falcons were still nesting on the sandstone banks of the North Saskatchewan River that had its headwaters in the Rocky Mountains, 300 km to the west of the city. But by 1967, the last of the falcon eyries in the southern half of the province stood empty, the result of environmental pollution superimposed on the robbing of all nest young by animal traders and pseudo falconers (Dekker 1967). Fortunately, arctic Peregrine populations continued to survive, and I discovered that they passed through central Alberta during their spring and fall migrations. An ideal place to study them was Beaverhills Lake, an extensive wetland one hour's drive east of the city. The lake's reedy shores and wide vistas reminded me of the former *Zuiderzee* in Holland. For nearly half a century now, Beaverhills Lake has been the centre of my intensive observations on the hunting habits of the Peregrine Falcon.

In modern times, the focus of biology students has largely shifted from outdoor to indoor studies, replacing the former field naturalist by a specialist in some laboratory. Be that as it may, in my view there is all the more opportunity for making discoveries the old-fashioned way. I like to think that I have been able to complement the work of the professional wildlife biologist who seldom gets the time for the non-intrusive study methods that come naturally to me.

"A naturalist," as recently defined by Schmidly (2005:452), "is the person who is inexhaustibly fascinated by biological diversity and who does not view organisms merely as models, or vehicles for theory, but rather as the thing itself that excites our admiration and our desire for knowledge."

CHAPTER 1

General introduction

This study began during an important era of environmental upheaval marked by widespread population declines of raptors and their reproductive failures in Europe as well as North America. The calamitous crash was eventually linked to the agricultural uses of organochlorine pesticides, including DDT and dieldrin (Ratcliffe 1963, 1967). The Peregrine (*Falco peregrinus*) proved to be especially vulnerable to the insidious effects of these synthetic poisons (Hickey 1969). The established knowledge of how the chlorinated hydrocarbons did their damage to the avian community was based on abundant evidence that sub-lethal residues caused eggshell thinning, which resulted in their breakage (Peakall et al. 1990). In addition, the most heavily polluted prey, as well as seed-eating birds that had been deliberately poisoned by agriculturalists, were probably first to fall prey to opportunistic predators including the falcons (Ratcliffe 1993). The subsequent recovery of the Peregrine falcon, now a matter of record, was associated with reduced pesticide levels in eggs and body tissue during the 1980s (Cade et al 1988; Court et al 1990).

The toxic chemical threat caused deep concern about the future of the Peregrine and sparked the 1965 International Madison Peregrine Conference (Hickey 1969). While the main debate about the decline was centred in Europe and eastern North America, the thinly populated Rocky Mountain region did not escape from the harmful spread of industrial pesticides. In the mid 1960s, an extensive survey of the western United States and Canada found less than 30% of known traditional breeding sites still occupied (Enderson 1965). The last of the nests in agricultural regions of southern and central Alberta became vacant by the late 1960s (Dekker 1967; Fyfe 1976). Subsequent checks by federal and provincial biologists found toxic residues in Peregrine eggs in northern Alberta and even on the remote coast of the Arctic Ocean as late as the 1980s (Court et al. 1990; Court 1993).

Meanwhile, ideas formulated during the landmark Madison Conference and the subsequent Peregrine Conference in Sacramento (Cade et al. 1988) led to innovative efforts to restore the Peregrine through captive breeding and reintroductions. Coupled with the efforts by others to reduce and eventually ban the use of certain toxic pesticides, the breeding program became an unqualified success thanks to the expertise and dedicated efforts of falconers. The releases of juvenile falcons proved crucial to the re-establishment of the endangered Peregrine in regions where the species had become extinct. By the beginning of this century, Peregrines were well on the way to a near-complete recovery in all of North America (Cade and Burnham 2003; Holroyd 2003), including the Canadian provinces of Alberta and British Columbia, which are the location for the field studies described in this dissertation.

Against the above-described historical background, I was motivated to begin my observations by the need for first-hand information on wild falcons. At that time, now half a century ago, the existing reference books contained a wealth of detail on Peregrine distribution, subspecies, plumages, morphology, and nesting habits (Fischer 1967; Brown and Amadon 1968). As well, food habits had been extensively studied across the falcon's worldwide range, but mainly through the collection of prey remains at nest sites. However, information on its hunting habits was primarily derived from falconry birds, which make their kills under contrived circumstances arranged by the falconer. By contrast, published reports of wild Peregrines in the act of capturing prey were rare and anecdotal (e.g. Beebe 1960; Bent 1961; Herbert and Herbert 1965; Fischer 1967). An exception was the paper by Gustaf Rudebeck (1950), a Swedish ornithologist, who watched migrating raptors at Falsterbo on the south coast of Scandinavia. By way of an explanation for the scarcity of detailed information, the British authority on the species, Derek Ratcliffe (1980:128) argued that "the study of food by direct observation of Peregrines in the act of killing prey is not really a practical proposition.... a vast amount of time would be needed to collect a reasonable sample of observation."

If we accept Ratcliffe's claim, then this thesis must be judged an impractical proposition. Indeed, it has taken me an improbably long time to amass the data presented in these chapters. More than 40 years ago, when I first read the singular paper by Rudebeck (1950), I did not think that I would ever match his list of 19 captures observed. As Rudebeck pointed out, the foraging studies of a raptor should involve many different birds, since individuals can vary much in their choice of prey. Consequently, he argued, a study of migrating Peregrines provides a more balanced picture of the species' menu than a pair of adults at their nest site. This was an added inspiration to me during the early years when I began my observation in central Alberta at a time when the Peregrine was an increasingly rare and threatened species. By now, of course, the picture has completely changed for the better, and I eventually beat Rudebeck's score by a factor of 25. The slow rise of my tally of hunts and kills over time is shown in Figure 1.2 and Table 16.1.

From the outset, the objective of my field studies was to collect the largest data set possible on the hunting habits of the Peregrine, based on first-hand observation. Secondly, I was interested in how the prey, in particular shorebirds, attempted to evade capture. This too was an area about which little or nothing had been published at the time.

THE FALCONS

The main subject of this thesis is the **Peregrine Falcon** (*Falco peregrinus*), an aerial hunter of birds believed to be the fastest creature on earth, capable of diving speeds that have been measured at over 350 km/hour (Franklin 1999), and with one of the widest distributions of all avian species. Practically cosmopolitan, it occurs in suitable habitats on all continents with the exception of high arctic tundra, deserts, Iceland, and New Zealand (Brown and Amadon 1968). The nominate form (*F. p. peregrinus*) was described for Britain (Ratcliffe 1980). Worldwide up to 19 subspecies or geographic races have been recognized, three of which breed in North America: *F. p. anatum, F. p. pealei*, and *F. p. tundrius* (White et al. 2002). Between them, there are subtle differences in plumage and wing proportions, and their typical ranges are widely separated. The anatum subspecies – formerly known as the Duck Hawk – breeds across the continent; the Peale's falcon is restricted to the Northwest Pacific coast; and *F. p. tundrius* migrates twice yearly between its arctic nesting grounds and tropical wintering ranges. Breeding farthest north, *tundrius* is the smallest North American Peregrine and lightest in colour, which contrasts with Europe, where the biggest and palest Peregrines (*F. p. calidus*) are found at the highest latitudes.

Subjects of secondary prominence in this study are three other falcons, all of which, like the Peregrine, prey on birds. The **Merlin** (*Falco columbarius*) occurs in Europe as well as North America, but the physical differences between them are very minor. The pale *F. c. richardsoni* of central Alberta is somewhat larger than its European conspecifics (Brown and Amadon 1968), but its hunting habits are much the same.

In the past, the circumpolar **Gyrfalcons** (*Falco rusticolus*) were split up into several subspecies or geographic races, but today they are considered one and the same. Both in Eurasia and America individuals vary in colour from nearly white to almost black (Cade at al. 1998; Potapov and Sale 2005).

The **Prairie Falcon** (*Falco mexicanus*) is indigenous to the Americas. It is about the same size as the Peregrine, which occurs mainly near water, but the Prairie Falcon is a dry-country raptor that preys on small mammals as well as on birds, similar to Lanner (*Falco biarmicus*) and Saker Falcons (*Falco cherrug*). In Alberta, the Prairie Falcon is a year-round resident as far north as the latitude of Edmonton (Dekker and Corrigan 2006), whereas the Peregrine is highly migratory. On sympatric breeding range along the rivers that transect the semi-arid plains of central and southern Alberta, the Prairie Falcon competes with the Peregrine for nesting sites.

An early research question I posed, having to do with the pesticide threat and the alarming reports of eggshell thinning and breeding failures, was this: Were northern Peregrines that migrated through Alberta still reproducing at normal levels? And what was the proportion of juveniles in the cohort? Reproductive output is a critical parameter of the health of wildlife population, and Peregrine mortality during the first winter is believed to be high (White et al. 2002). How many first-year falcons survived to make the return migration to the Canadian arctic? No such information was available when I began my 15-year survey of migrating falcons at Beaverhills Lake in central Alberta (Chapter 2).

The decision to include four species of falcons in this thesis was inspired by the opportunities encountered during 48 years in the field, enabling me to compare the success rates of the Peregrine with that of the Merlin (*Falco columbarius*), which also hunted shorebirds at Beaverhills Lake (Chapter 11). The bird-hunting tactics of the Prairie Falcon (*Falco mexicanus*) and the Gyrfalcon (*Falco rusticolus*) were poorly known. Descriptions of their food habits were mainly based on the collection of prey remains at nest sites, as was the case with the Peregrine. The unique chance to compare the methods and success rates of the Prairie Falcon and the Gyrfalcon came about after I had suffered a back injury that prevented me from walking. All I could do at that time was sit in the car and wait for raptor attacks on a flock of feral pigeons at an industrial site in the city of Edmonton. Comparative information on bird-hunting tactics of these two falcons had never before been published. Furthermore, this thesis also details intra- and interspecific klepto-parasitic incidents between all four species, as well as their interaction with other raptors, in particular the American Bald Eagle (*Haliaetus leucocephalus*).

STUDY AREAS

The study areas include five different locations in Alberta, Canada, three in coastal British Columbia, Canada, (Figure 1.1), and one in the Netherlands.

(**1**) 1960–2008. The valley of the Red Deer River in central Alberta. There, Peregrines and Prairie Falcons compete for nesting cliffs and prey.

(**2**) 1965–2008. Beaverhills Lake, a 140 km^2 Ramsar wetland east of Alberta's capital city Edmonton. The lake is surrounded by agricultural fields, rough pastures, and woodlots. Migrating Peregrines and Merlins occur during spring and fall. Merlins also breed in the region. Both species hunt migrating shorebirds.

(**3**) 1980–1994. A 5-km^2 enclave of low-lying agricultural fields on mountainous Vancouver Island in British Columbia. Wintering Peregrines hunt mainly ducks and are frequently harassed by Bald Eagles.

(**4**) 1994–2008. Boundary Bay, a 15-km section of coastline and intertidal mud flats near the city of Vancouver. It is a traditional wintering range for Dunlins (*Calidris alpina*), which are preyed upon by Peregrines and Merlins.

(**5**) 1995–1996. Langara Island, a heavily forested island well off the coast of British Columbia and just south of Alaska. Locally breeding Peregrines hunted mainly seabirds over the ocean.

(**6**) 1998–2008. Wabamun Lake, 60 km west of Edmonton. Over ten years, a pair of Peregrines has nested in a box attached to a lakeside powerplant. The falcons often hunted cooperatively and preyed largely on gulls. Also detailed in this chapter are the interactions of the parent Peregrines and their fledglings, with particular reference to the question of whether the adults teach their young how to hunt.

(**7**) 1998–2000. The inner city of Edmonton. Over two winters, feral Rock Doves (*Columba livia*), frequenting a grain-loading railway siding, were hunted by Prairie Falcons and Gyrfalcons. Their respective capture methods and success rates are described and compared.

(**8**) 1999–2002. A section of the North Saskatchewan River flowing through an agricultural region of central Alberta. Along a stretch of water kept ice-free by city effluents, wintering Gyrfalcons and Bald Eagles hunted Mallards (*Anas platyrhynchos*) flying inland to forage on farms.

(**9**) 2006–2008. (Not indicated on map below). The coast of the Dutch Wadden Sea in Friesland. Here, migrating and wintering Peregrines captured a range of prey including Dunlins.

For further details of the study areas, see chapters 2 to 15.

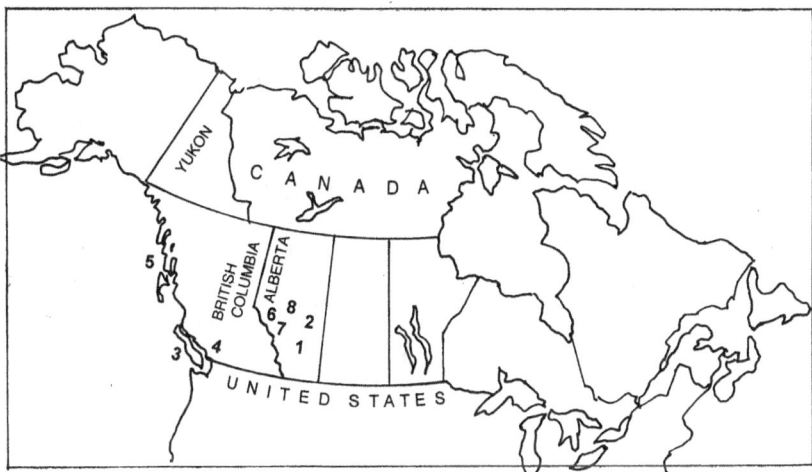

METHODS

My methods of finding falcons were hands-off and non-intrusive, requiring only patience and keeping my distance. The less disturbance I caused in the field, the better my chances of seeing predator-prey interactions unaffected by the human presence. My main problem was to decide on the question of what constitutes a hunt by a Peregrine, and what should be considered only play or harassment of other birds. This conundrum was discussed at length in the literature (Brown and Amadon 1968; Ratcliffe 1980; Treleaven 1980). I too struggled with the question of correct interpretation and decided that the only criterion to go by was whether the attack had been completed and the result became known. In this thesis, the terms hunt and attack are used interchangeably for one attempt with known outcome by a falcon at capturing a bird of a species commonly preyed upon. One hunt or attack could include more than one attempt at grabbing that prey. I did not count as hunts any aggressive-looking passes at large birds such as geese, swans, or buteo hawks. Neither was the capture of insects and rodents included in the tables of hunts and kills.

A pass at a flying crow that evades the falcon by a wide margin and is not pursued further could not be considered a serious attempt at capture. On the other hand, any close pass at a bird of a size known to be preyed upon by Peregrines was classified as a hunt even though the target might escape capture or was let go. Over the years, I became convinced of the accuracy of this interpretation. I learned that a hunting adult Peregrine can make a deceptively half-hearted impression, whereas immatures might harass birds vigorously soon after they had eaten their fill. The question of deciding what constitutes a hunt and what not became further confused after I saw falcons grab flying ducks and let them drop into the reeds, or carry them down to the ground but leave them uneaten.

Understanding a creature like the Peregrine may largely lie outside our realm. Perhaps it's easier to empathise with the prey, for humans are familiar with fear. A prey under attack can choose whether or not to react in a timely evasive way. Birds that make a wrong move may end up in the falcon's clutches whether it is hunting in earnest or just playing. In their hurry to get out to the way, some prey are unlucky and hit a barbed-wire fence or overhead wire. On the Pacific coast, I twice saw a Dunlin drop out of a dense flock careening low over the water in panicked agitation, probably induced by fear of falcon attack, but there was no raptor to be seen. It is a moot question, whether these Dunlins had collided with their fellows or perhaps suffered a coronary failure; evidently they were victims of predation in an indirect but fatal way. Both of these Dunlins, splashing as if lifeless into the shallows, were quickly grabbed by Bald Eagles.

The hit or miss moment of a falcon's attack is a split-second affair easily misinterpreted by the human observer. Birdwatchers and waterfowl hunters have told me of seeing falcons fly along a shore and "knock one duck dead after the other, and just letting them drop into the water." Even geese that awkwardly landed or fell down to evade an aggressive Peregrine were believed to have been struck and wounded. Similar reports are contained in Bent (1938). These stories may well have been inspired by the common descriptions in the reference books to the effect that Peregrines hit their prey in mid-air, so that the victim falls to the ground dead or crippled, sometimes with its head cut off by the aerial strike (Ratcliffe 1980). Such events may not be unusual for trained Peregrines that "wait on" high above a falconer until the dogs flush a crouching grouse or partridge from the ground below the falcon. However, a wild Peregrine will smack other creatures only in defence, for instance at the nest site. When foraging for food, it commonly secures the prey in its feet and carries it along. Allowing a wounded bird to fall might come at the risk of losing it in vegetation. In the few cases in which I have seen a wild falcon hit a bird in the air, sending feathers flying, it looked to me like a failed attempt at grabbing that prey. Some birds, such as Rock Doves and grouse, have loose feathering that easily separates in the claws of a raptor.

There is a simple explanation for the fact that aerial hits can be misinterpreted. Flying ducks and other prey species routinely plunge to the ground or into water to dodge an attacking falcon. The timing of these evasive tactics is so acute that it indeed may look as if the target was struck. For this reason, to decide whether a kill had actually taken place, I have relied on certainties, not assumptions. To be sure of a hit, I either had to see feathers fly when the bird fell down, or see the falcon descend on the spot, either to feed right there or carry the prey off to a distant post.

To determine hunting success rates of the falcons, I divided my observation of hunts (or attacks) into three categories: (A) hunts that took place under perfect viewing conditions and at once allowed me to see whether or not the intended prey was seized; (B) hunts that were partly obscured by distance or vegetation so that the outcome could only be determined subsequently, for instance, if the falcon was found feeding on a prey, or if it flew away to attack elsewhere; and (C) hunts that remained undecided and their results unknown.

This last category included a high percentage of spectacular stoops from a great height, and far-ranging descents at accelerating speed, as well as some very persistent pursuits of fleeing birds that carried on out of sight into the distance or high into the sky. In the Alberta study of migrating falcons, category C observations accounted for 30% of all hunts recorded. Although they contributed much to my appreciation and understanding of the Peregrine as a hunter, they could not be used in calculating hunting success rates. As to the difference between A and B categories, after 15 years of detailed data collection, there was no significant difference (x^2 chi-square = 0.0039, 1df NS) in their respective success rates of 7.4% and 8.3%. Henceforth, I have combined both categories in my classification of hunts and kills.

Based on experience, I eventually developed some simple strategies to enlarge my chances of observing hunts. The first priority was to select an observation point near a wetland or section of coastline that combined an unobstructed view with the possibility of approaching it without disturbing nearby ducks or shorebirds. I usually sat down or used the parked car as a comfortable blind. Wind direction and sun angle were also of critical importance. However, one of the most unpredictable yet decisive factors was plain good luck to be in the right place at the right time.

My eyesight is by no means good, even quite poor under low light conditions. By way of compensation I continually scan the land or water through wide-angle binoculars. Alerted by their alarm calls and sudden flushing, I take my clues from the prey species. Unfortunately, it is still possible, even likely, to miss the action and only discover the falcon after its strike is over. At close range, a

Peregrine is easily missed. Spotting distant falcons is easier. After noting a far-off rising of shorebirds, I scan the horizon and with luck pick up the Peregrine climbing away after an unsuccessful attack or descending for a new attempt. One of my early mistakes was to hurry to the distant section of shore where the disturbance had just taken place, but falcons seldom strike twice in the same spot after prey have been put on high alert. One low pass by a Peregrine makes ducks flee to the safety of deeper water. Waders often leave the locality. A hunting falcon causes a great deal of unrest among prey species, resulting in the desertion of wide stretches of coastline. By contrast, if a Peregrine has not shown up for some time, loafing ducks dawdle on the shore and sandpipers forage in the shallows without interruption. Noting the general calm, I purposely selected these places to sit down and wait. Frequently scanning the shoreline in either direction, my objective was to spot an approaching falcon before the shorebirds did. If an attack materialized, I could follow the entire hunting sequence from its very beginning. It often involved hours of waiting and strenuous effort of keeping the shaking binoculars focussed on a distant speck of a falcon soaring far inland or at heights estimated to be >1 km. Not all sightings, even those that began with much promise, led to the desired results. But by and large, my efforts proved effective over time.

A secondary strategy was to watch perched falcons, sitting on a fence post or fieldstone, until they flew off to start hunting. Finding them became easier over time as I learned their favourite places. However, staring at an inactive falcon is a very boring business that may take hours and still end in failure if I looked away for a minute, or when the falcon eventually flew out of sight. It was prudent not to approach too closely, and resist the desire to see plumage details. Falcons differ in their tolerance of humans. The best thing to do is to keep one's distance so as not to influence their behaviour. Most times I was unable to see the perched falcon with the naked eye, forcing me the keep the binoculars trained for long periods or to squint through the telescope. Providing the falcon eventually became active and decided to fly into my direction, the chances of seeing an interesting hunt were best the greater the initial distance between the bird and my point of observation. A hunting falcon, looking far ahead, often takes a long-range approach before reaching top speed. Of course, just as often it dwindled away into the distance.

The above methods pertain to the difficulties of watching migrating and wintering falcons. Observation can be much more productive near nest sites because territorial adults use the same high perches and begin foraging flights at predictable times of day. These conditions applied to a pair of Peregrines breeding in a nest box attached to a tall chimney of an electricity generating station at Wabamun Lake in Alberta. By contrast, however, at the cliff nesting sites of Peregrines along the Red Deer River, opportunities for recording hunting behaviour were non-existent. There, the falcons did all of their hunting over the adjacent fields out of sight of my observation point in the narrow river valley. They simply flew away or soared to a great height until the trees cut them off from view.

PROJECT OUTLINE AND OBJECTIVES

The above methods were applied to the following sequence of projects and the data collected form the basis for this thesis.

(1) After the Peregrine breeding population of central Alberta had died out, their return became of special concern to provincial and federal government agencies who eventually, with the financial support of a major oil company, organized the release of captive-raised Peregrines along the Red Deer River where the species formerly nested. As a volunteer familiar with the pre-release situation, I assisted wildlife biologists in monitoring the Peregrine's success in returning to the river. My inventory surveys, which began in 1960, continue as of 2008. A unique part of these investigations was the interspecific competition, for nest sites and prey, between Peregrines and Prairie Falcons (Chapter 12).

(2) After the Peregrine had become extinct as a breeding bird in the lower half of the province, I discovered that falcons of unknown origin were still passing through the province in spring and fall. When I notified the Alberta Fish and Wildlife Division, I was hired on a volunteer basis to monitor the migrations at Beaverhills Lake in central Alberta. At that time nothing was known about the spring passage of Peregrines in North America, while fall flights had been continuously monitored on islands off the Atlantic seaboard and along the Texas Gulf coast. In addition to timing the migrations, I kept track of the adult versus immature ratios as a useful parameter of the population dynamics and reproductive health of the migrants. I also collected the remains of birds killed by Peregrines in order to have them analyzed for pesticide residues by an Alberta government laboratory. The program was conducted for 15 years under contract to the Alberta Fish & Wildlife Division, but I continued on my own after 1980, albeit in a less focussed manner, until 2008 (Chapter 2).

(3) Apart from monitoring the spring and fall migrations, my major interest was the foraging habits of wild falcons at Beaverhills Lake. It led to the publication of a very large data set on hunts and kills in 1980 and 1988 (Chapters 3 and 11).

(4) Alberta Peregrines are highly migratory and few if any remain in the Edmonton region after the first week of October. With the objective of studying wintering Peregrines, I travelled in January of each year from 1980 to 1994 to Vancouver Island on the west coast of British Columbia. At the time, little or nothing was known about the food habits of these falcons and their competitive relationship with the Bald Eagle (Chapters 4 and 5).

(5) After having attained an understanding of the Peregrine's winter ecology in the agricultural fields of Vancouver Island, my curiosity led me on to Boundary Bay, a section of the Fraser River Delta on the coast of Pacific Ocean near Vancouver. There, I studied Peregrine predation on Dunlins and ducks. After several years of data collection, I cooperated with researchers from the Canadian Wildlife Service and Simon Fraser University, who had intensively studied shorebird migrations in the delta for decades. A major focus was on the relationship of Peregrine predation rates on Dunlins with the tidal cycle (Chapters 7 to 9).

(**6**) In 1995, I obtained an invitation from Peregrine researcher Wayne Nelson to accompany him on a nesting survey of Langara Island in the Pacific Ocean, just south of Alaska and west of the mountainous coast of British Columbia. It led to further observation in the following year and unique discoveries of the way these falcons hunted seabirds (Chapter 6).

(**7**) An opportunity to study the foraging behaviour of breeding Peregrines arrived after reintroduced falcons began using a nest box attached to a high chimney of a coal-fired electrical generating station at Wabamun Lake in central Alberta. There, I began intensive observations in 1998, and by 2004 I had collected a large sample of hunts and kills. An additional area of interest was parent-fledgling interaction. Here too, my observations were carried on after the publication of a 2005 paper on seven years of data collecting (Chapter 10).

(**8**) Based on long-term research at Beaverhills Lake, I compared the success rates of the Peregrine with that of the Merlin on the same kind of prey, namely small shorebirds and passerines (Chapter 11).

(**9**) Klepto-parasitic interaction between Bald Eagles and Peregrines became part of the investigations at Boundary Bay, Vancouver Island, and Langara Island (Chapters 5 to 9).

(**10**) An opportunity to observe the various hunting methods and success rates of the Prairie Falcon and the Gyrfalcon arose in the winters of 1998–2000 when both species preyed on feral Rock Doves in the city of Edmonton. At the time, the existing literature contained very little detail about the bird-hunting habits of either of these falcons (Chapter 13).

(**11**) During the winters of 1999–2002, I studied the hunting methods of Gyrfalcons preying on Mallards wintering along an open stretch of river downstream from the city of Edmonton. An interesting factor in the predator-prey dynamics was klepto-parasitic interference from Bald Eagles. Prior to these studies, published information of the Gyrfalcon's food base was mainly restricted to prey remains collected at nest sites (Chapter 14).

(**12**) In the late fall of 2007 and 2008, I spent three months in the Netherlands to study the foraging habits of migrating and wintering Peregrines on the coast of the Wadden Sea with special reference to Dunlin anti-predation behaviour. In particular, my objective was to determine whether or not the Dunlins wintering on the Dutch coast would engage in over-ocean flocking during high tides when all intertidal habitats were inundated (Chapter 15).

The above studies and data were published over a period of 29 years in six different refereed journals. In the Discussion and Synthesis (Chapter 16) I provide an overview of the major findings.

Figure 1.2. The graphs summarize the slow accumulation of data on the hunting habits of Peregrines, substantiating Ratcliffe's (1980:128) claim to the effect that "a vast amount of time would be needed to collect a reasonable sample of observations." The data points represent the number of hunts and kills reported in my major papers in order of their publication dates (Chapters 3, 5, 6, 8, 10, 15, and additional unpublished data). Depending on location and season, between 1.5 and 5.5 hours of field time were on average required to observe one hunt, with an overall chance of about one in ten hunts ending in a capture. The jogs in the line are caused by variations in the results obtained between locations and seasons. The slow rise in hunts and kills recorded during the first three decades, as compared to the steeper annual increases thereafter, can be explained as follows. During the early years I exclusively watched migrating Peregrines in Alberta, which have a lower success rate than wintering or nesting falcons. As indicated by the graph, records began to increase when I added the observations of falcons wintering in British Columbia, and the line reached its steepest pitch after 2004 with the inclusion of the records of a breeding pair at Alberta's Wabamun Lake.

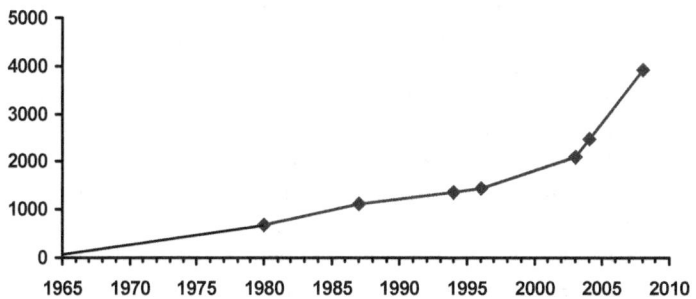

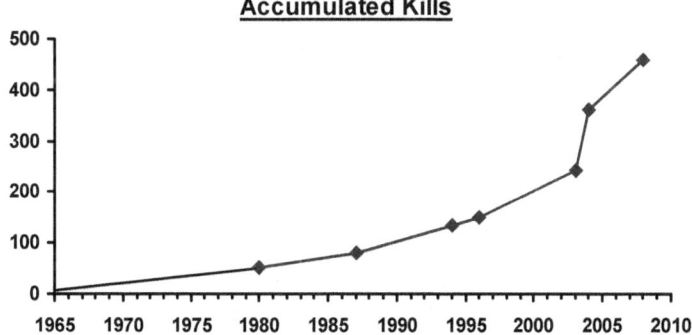

CHAPTER 2

Peregrine migrations in Alberta

Dekker, D. 1979. Characteristics of Peregrine Falcons migrating through central Alberta, 1969–1978. Canadian Field-Naturalist 93:296–302.

Dekker, D. 1984. Spring and fall migrations of Peregrine Falcons in central Alberta, 1979–1983, with comparisons to 1969–1978. The Journal of Raptor Research 18:92–97.

Abstract – During the spring migration periods of 1969–1983, I recorded a total of 880 sightings of Peregrine Falcons at Beaverhills Lake in central Alberta. Fall sightings were far fewer ($n = 51$), suggesting that the southward migration follows a different route than the northward migration. Earliest spring dates and last dates were respectively 16 April and 31 May. The majority (82%) of falcons were seen between 4 May and 23 May. A comparison of 5-year means indicates that the timing of the spring migrations stayed the same over the 15-year period. First arrival dates, mid dates, and last dates for adults were 27 April, 8 May, and 20 May, respectively nine, seven, and five days ahead of immature falcons. Mean sightings per hour were 0.22 in 1974–1978 and 0.23 in 1979–1983, suggesting that there was little or no change in the number of falcons passing through the study area. The ratio of adults to immatures was roughly 1:1 in spring and 1:2 in fall. The sex ratio was unbalanced, with females outnumbering males by a factor of three to one. Some Peregrines looked like *F. p. anatum*, others like *F. p. tundrius*, but the vast majority of sightings could not be assigned to either subspecies. Falcons that departed the study area commonly soared to high altitudes, and strong head winds did not deter adults from resuming their northward migration. Foraging activity took place throughout the day and as late as one hour after sundown.

INTRODUCTION

During the latter half of the 1900s, the Peregrine Falcon (*Falco peregrinus*) declined in all settled regions of North America due to toxic residues of chemicals used in agriculture and forestry (Hickey 1969). As a breeding bird it became extirpated in central and southern Alberta by the early 1970s (Dekker 1967; Fyfe et al. 1976). Arctic and sub-arctic populations survived and their fall migrations were monitored at traditional flyways on the Atlantic coast, the Gulf of Mexico, and the Great Lakes (Mueller and Berger 1961; Ward and Berry 1972; Hunt et al. 1975). However, at that time very little was known about the northward flight. Isolated spring sightings of Peregrines were reported from across the continent, as evidenced by scattered notes in *American Birds*. During a spring survey of migrating raptors along Lake Ontario, only four Peregrines were seen in 108 days (Haugh and Cade 1966). The paucity of data led to speculation about the timing, routes, and age structure of Peregrines migrating northward. The fact that few sub-adults were reported on the breeding grounds in Alaska appeared to confirm that mortality of first-year falcons is high. Cade (1960) postulated that some immature Peregrines do not return to the arctic and sub-arctic until their second year. This paper reports on Peregrines migrating through central Alberta, Canada, in fall as well as spring, over 15 consecutive years from 1969 to 1983.

STUDY AREA AND METHODS

The study area is at Beaverhills Lake, a 10 x 18 km sheet of shallow water. The latitude is 53° 20', and the local climate is cold continental. From early November to late April, the lake is frozen and snow-covered. Water levels are dynamic, varying with annual precipitation, and the littoral zone is subject to periodic changes in plant succession. Shoreline habitat includes mud flats, reed beds, rough cattle pastures, and copses of trembling aspen (*Populus tremuloides*) and willow (*Salix* spp.). In spring and fall, the lake is a traditional staging site for migrating geese, ducks, gulls, and shorebirds. The most common shorebird species, that periodically occur in large numbers, include Black-bellied Plover (*Pluvialis squatarola*), Semipalmated Plover (*Charadrius semipalmatus*), Lesser Yellowlegs (*Tringa flavipes*), Semipalmated Sandpiper (*Calidris pusilla*), Least Sandpiper (*Calidris minutilla*), Baird's Sandpiper (*Calidris bairdii*), Pectoral Sandpiper (*Calidris melanotos*), Short-billed Dowitcher (*Limnodromus griseus*) and Red Phalarope (*Phalaropus fulicaria*). The shorebirds are hunted by Peregrine Falcons that shadow their prey on its continent-spanning migrations. For a detailed description of the local avifauna, see Dekker (1991, 1998).

I visited the study area at irregular intervals from early spring to freeze-up, but mostly during the migration periods 15 April to 31 May, and 1 September to 15 October. Respectively, the seasonal total of visits was 446 and 223, with a mean of 30 days in spring and 15 in fall. Field days varied in duration, some lasting from sunrise to nightfall; others spanning 3–4 hours within varying time frames. During the peak of the spring migration, I often stayed overnight in the area. Hours afield were counted during 1974–1983, but not in 1969–1973. The accumulated total number of hours was 3,068 with a mean of 9.5 hours/day. The time of day when Peregrines were sighted was only recorded during the last five years.

In my search for Peregrines, I followed no standard procedures but my methods were similar over 15 years. Each day I walked 5–25 km, frequently pausing to scan the area through binoculars or a telescope. During wet, windy, or hot weather, I spent many hours sitting in a car parked at locations overlooking the lakeshore. Alarm calls and the sudden flushing of shorebirds alerted me to the approach of Peregrines. Flying falcons were followed in the glasses for as long as they remained in view. Some falcons that perched on the ground or fence posts were watched until they left of their own accord. About 85% of falcons sighted could be positively identified as Peregrines. During spring, distant large falcons were assumed to be Peregrines, unless I was aware of the local occurrence of a Prairie Falcon (*Falco mexicanus*) or Gyrfalcon (*Falco rusticolus*). Both species have been recorded in the area, but seldom during the spring period. In late summer and fall, when Peregrines are less common, unidentified distant falcons were deleted from the tally. For criteria used in field identification of large falcons, see Dekker 1977, 1982.

Repeat sightings of the same Peregrines could lead to misleading totals. To deal with the problem, I initially kept separate lists for maximum and minimum sightings. The duplication rate during the first 10 years was two to one, indicative of the magnitude of the problem (Dekker 1979).

However, for the last five years, my daily tally included all sightings except known and suspected duplications. Under favourable light conditions, Peregrines can be distinguished from each other by differences in plumage, especially when seen together. Peregrines attract each other, and their migration is not an evenly spaced event. There were days when none were seen, followed by brief periods when two or more falcons passed by simultaneously or in quick succession. I made no attempts at trapping them, and I kept my distance from perched falcons so as not to disturb their normal behaviour.

In good light, adult Peregrines can be distinguished from first-year immatures by plumage characteristics (Brown and Amadon 1968; Palmer 1988). Adults are dorsally blue-grey or slate-grey; immatures a shade of brown. Differentiating between the sexes is difficult, although sexual size dimorphism in Peregrines is pronounced, and the smallest males are about two-thirds the size of the largest females. However, size is a relative and deceptive criterion under field conditions. In assigning sex, I have limited myself to cases when I could do so with confidence. Observations were written down during or at the end of the day.

RESULTS AND DISCUSSION

Number of sightings and timing of the migrations. In spring, the first Peregrines to arrive in the study area were always adults (Figure 2.1). The earliest sightings ranged from 16 April to 30 April. The 15-year mean of first arrival was 26 April for adults, nine days ahead of the immatures. The mid dates, when 50% of birds had passed, were 8 May for adults and 15 May for immatures. The mean last dates were respectively 20 and 25 May (Table 2.1). Stragglers were seen until the last day of May and into early June. Immature falcons were in the majority with 38% of all sightings, compared to 32% for adults, but 30% of sightings could not be classified as to age group (Table 2.2). Although I was unable to reliably tell the sex of 54% of all sightings, based on perceived differences in size, female Peregrines appeared to outnumber males by a factor of three to one. Females were also more numerous than males in coastal migration surveys (Ward and Berry 1972; Hunt et al. 1975).

Sightings per year ranged widely from 15 to 165, with the lowest 5-year mean of 27 in 1969–1973, as compared to 81 in 1974–1978, and 68 in 1979–1983 (Table 2.3). The comparatively low results of the first five years may be due to less expertise in finding falcons, or less hours spent in the field at the most productive morning hours. Peregrines were seen throughout the day, but sightings/hour were highest before 12:00 and lowest in the afternoon (Table 2.4).

Hours afield were only recorded in the last ten years. Sightings per hour varied from a low of 0.14 in 1975 to a high of 0.43 in 1977 (Table 2.4). The 5-year mean of 1979–1983 was 0.23/hour, which is lower than the 0.26/hour of the previous five years, but the difference was not statistically significant ($P > 0.05$). Furthermore, by a downward adjustment of the outlier in 1977, the 5-year mean evens out to 0.22, nearly equivalent to the 0.23 of the previous five years. The statistical line graph is flat, with a regression coefficient of $r^2 = 0.002$ (Figure 2.2). This suggests that the number of migrants has remained virtually the same over these ten years.

The exceptionally high number of sightings in 1977 can be explained on the basis of climatic conditions. That year, a severe regional drought had resulted in the drying up of wetlands in the surrounding countryside, which led to a concentration of migrating shorebirds on the lakeshore. By the same token, receding water levels had created wide margins of mudflats, attracting huge flocks of sandpipers and many Peregrines. On some days I recorded more than 20 sightings. By contrast, the lowest sightings/hour were recorded in the wet years of 1974 and 1975. Then, record high lake levels had inundated all mud flat habitats, and the shorebirds were dispersed over the surrounding countryside, where snowmelt had collected in numerous depressions, creating a mosaic of seasonal ponds and sloughs.

Although the number of fall sightings is too small for a meaningful comparison between the years, the number of sightings/day increased from 0.24/day in 1969–1978 to 0.30/day in 1979–1983. Why Peregrines are much scarcer in fall than in spring is not clear. The most likely reason is that most northern falcons travel southbound along a different route, away from Alberta. This would explain their concentrations at known fall migration points along the Atlantic coast (Ward and Berry 1972). By contrast, on their way back to arctic breeding grounds, falcons apparently choose a more direct route through the centre of the continent. A differentiation of spring and fall migration routes also has been documented for some shorebird species (Godfrey 1966).

Subspecies and plumage variation. The North American population of the Peregrine Falcon includes three subspecies (Palmer 1988). *F. p. anatum* occurs continent-wide in suitable habitat south of the tundra. The largest sub-species is *F. p. pealei*, which inhabits the Pacific coast of northwestern Canada and Alaska

(Beebe 1960). The smallest and relatively long-winged *F. p. tundrius* breeds farthest north and travels twice yearly between its arctic nesting grounds and its South American winter range (White 1968).

During this study, I saw no Peregrines that might belong to the coastal subspecies *F. p. pealei*. Most falcons loosely resembled the official plumage descriptions for *F. p. anatum*. Smaller falcons, perhaps representative of *F. p. tundrius*, were among the last to pass through. The standard I used to differentiate between them was dorsal colouration, which could best be appraised when the birds were in flight with wings spread. The blue-grey dorsal surface of a typical *F. p. anatum* is lightest on the rump and darkens to nearly black on the head, tail, and wing primaries (Beebe 1974). By comparison, when in flight, adults of *F. p. tundrius* are evenly blue-grey on the entire dorsal surface, and they look smaller and more slender than *F. p. anatum*. However, there is a lot of individual variation, and most of the migrants I saw could not be assigned with certainty to either one of the official subspecies. The presumed differences between them are no doubt diluted through intergradation as well as individual characteristics. Most notably, among the falcons that I saw were many dark, slender-winged birds that passed through in mid season. Dorsally, the adults looked almost black; the immatures a slate brown. These falcons might belong to an as yet unrecognized geographic race, possibly originating from the boreal forest region of northern Canada, south of the range of *tundrius*, and north of *anatum*.

All adult falcons, observed through a telescope or binoculars, had white chests, some with a pinkish overtone. The belly and flanks were barred or spotted, varying in density. Greater diversity of pigmentation and wider extremes in light and dark types existed among the first-year birds. In fall, immatures were darkly streaked on the entire under surface, but in spring, the majority had light-coloured, unmarked chests, resembling the adults. Apparently, these falcons had started their first feather moult during winter, and their plumage was bleached by tropical sunlight. To correctly classify these falcons, I needed to see their dorsal colour, which could range from blackish-brown to sandy. Some immature falcons, possibly of the *F. p. tundrius* subspecies, looked pale enough to resemble Prairie Falcons. Second-year Peregrines, that had already attained adult markings ventrally, might still retain some brownish feathers on their back and in the wings. Of interest was the presence of "red-tailed" Peregrines, which featured a reddish brown cast to the upper tail. Their unknown breeding range could well be a very restricted locality in the vast expanse of northern Canada.

Age ratios. The small sample of fall data I obtained included 26 immatures and 13 adults, a ratio of 2:1. Fall surveys from coastal migration routes reported even greater differences between the age classes, varying from 4:1 to 6:1 (Berry 1971; Ward and Berry 1972; Hunt et al. 1975). This highly unbalanced age ratio cannot be considered representation of the population as a whole. Shor (1970) calculated that Peregrine populations always contain more adults than first-year individuals, even at the time of the fall migration. This difference should be all the more pronounced in spring. Band returns indicate a high (70%) mortality among juvenile falcons during their first winter (Enderson 1969). So why the preponderance of immature falcons in my spring sightings?

Berry (1971) proposed that the uneven age ratios found during fall migration suggest differential migratory behaviour among the age classes. Instead of following the coastlines like the immatures tend to do, adult Peregrines might fly on a more direct, overland course to their wintering grounds. As experienced hunters, they need not depend on the open beaches, where migrating forest birds are relatively easily caught, even for first-year falcons (Hunt et al. 1975).

This scenario should not be applicable to the spring migrations. After having spent winter on their own, first-year Peregrines hunt the same prey and use the same hunting methods as adults (Dekker 1980). Hence the observed majority of immatures in my spring sightings cannot be explained on the basis of differing foraging ability or behaviour. There is another, more likely reason. On their way north, adult falcons are in a hurry to return to their breeding grounds, where nest site competition is fierce, and they are probably less inclined to stop over on attractive feeding grounds than are young falcons not yet of breeding age.

Sex ratios. As stated earlier, separating males from females on the basis of size impressions in the field is at best a guessing game. However, I had no reason to think that adult males were less common than adult females. This contrasts with migration counts from elsewhere. Adult male Peregrines are reported to be most elusive along the major autumn flyways along the Atlantic and Texas coasts, where falcons are captured for banding. Only eight out of 639 Peregrines trapped on Assateague Island from 1939 to 1971 were adult males (Ward and Berry 1972). And no adult males were among 250 Peregrines banded on the Texas beaches from 1952 to 1973 (Hunt et al. 1975). Percentages of immature males trapped along the major coastal routes were 30% (Ward and Berry 1972), and 23–30% (Hunt et al. 1975).

Hunt et al (1975) claim that there is no obvious reason why sex ratios should be unbalanced. The preponderance of females in the Texas surveys may reflect differential trap response and/or prey preference. Female falcons readily take pigeons, whereas males may prefer small passerine birds, which are more commonly found inland, away from the open coast. In my opinion, there is a seconddary reason. Aggressive females may force the smaller males away from places where Peregrines congregate. I saw females chase and rob males that were carrying prey. Other males transported their prey far inland before they landed to eat, or they soared to a great height and fed while on the wing. By contrast, females routinely consumed their prey at the site of capture.

Behaviour. The influence of atmospheric conditions on the movements of migrating raptors has been discussed at length by Mueller and Berger (1961), and Haugh and Cade (1966). They concluded that the majority of raptorial birds migrate when tail winds are blowing. Ward and Berry (1972) and Hunt et al. (1975) correlated highest numbers of Peregrines in fall with days of heavy cloud, light tail winds, and low temperatures. Similar weather condition usually accompanied the influx of falcons into my study area, in spring as well as fall. Newly arriving falcons were easy to spot, flying low along the lakeshore and perching on fence posts. Most arrivals took place in the late afternoon on overcast days when dropping temperatures heralded a change in weather, followed by rain. Some falcons stayed in the area for several days until the skies cleared.

Mueller and Berger (1961) remarked that hawks have rarely been observed in the process of departing on migration. In this regard, my observations appear to have filled a gap in knowledge. During the morning hours, I kept many perched falcons under surveillance until they flew away of their own accord. Such departures commonly took place around 10:00 or 11:00, after the temperature had risen to allow for the formation of rising air currents. Spring migrants left in a northwesterly direction; fall migrants flew south.

Fischer (1967) stated that Peregrines migrate close to the ground. Brown and Amadon (1968:65) wrote that falcons "usually fly rather than sail or soar." However, the Peregrines that I watched leaving the study area climbed very high and commonly soared, not only during optimum conditions of sun and wind, but also under overcast skies, and even when light rain was falling. In this respect, the long-winged falcons seemed a match for the buoyancy of gulls. When circling overhead, some soaring falcons ascended until they were just visible in 10-power binoculars, their estimated altitude >2 km. Occasionally, I watched them enter the cumulus clouds that top thermal currents.

In spring, the predominant wind direction was from the north or northwest, which did not prevent the falcons from departing. Flying upwind for some distance, climbing steadily, they set their wings and soared. During their circling ascent they drifted downwind until they resumed a northerly direction and glided through the wind rapidly with wings flexed at a slight angle. Even head winds gusting to 50 km at ground level did not deter these falcons from high-altitude travel. All falcons seen to depart under these contrary wind conditions were adults.

During midday and bright weather conditions, Peregrines were largely absent from the area (Table 2.4), but at any time of day a falcon could stoop down from the skies to attack shorebirds, resting or feeding in the shallows along the lakeshore. Successful falcons, after consuming their kill, again soared up to a great height and sailed northward on slightly flexed wings.

Falcons that arrived in the area during the late afternoon might perch for several hours before starting foraging flights. Immature Peregrines were often seen chasing shorebirds in the evening. Two females that I had watched from late afternoon to nightfall began attacking ducks flying inland at dusk. Through binoculars and against a clear sky, the birds were just visible, silhouetted against a cloudless sky. Crepuscular foraging was also observed by Rudebeck (1950, 1951), Beebe (1960), Fischer (1967), and Clunie (1976). Late foraging may explain the relative scarcity of falcons on the Texas beaches at sundown, which prompted Enderson (1965) to speculate on the possibility of nocturnal migration flights. (Since then, Peregrines fitted with radio telemetry have indeed been tracked at night far out over the Atlantic Ocean (Cochran 1985). Making use of rising air currents over the still warm water, they traveled great distances before returning to the coast.)

To learn about their roosting habits, I watched four Peregrines, sitting on fence post or shoreline stones, until it was too dark to see them. Only one falcon was still on the same perch at first light. The others had evidently flown off, perhaps to hunt. Although I never saw direct evidence of hunting at dawn, it is a likely scenario, in particular for falcons that had been unable to secure an adequate meal the previous evening. Very early foraging may also explain the relative inactivity of most Peregrines after sunrise. During mid morning, before resuming migration, several falcons took a second meal from the partly consumed carcass of a duck lying on the grass nearby, possibly the remains of their prey caught the night before or at dawn. Other falcons that soared up by mid morning suddenly stooped at ducks flying below. Evidently, these Peregrines did their foraging along their way.

Table 2.1. Mean early dates, mid dates (half of all sightings), and mean late dates for adult (Ad.) and immature (Imm.) Peregrines sighted during spring migration in central Alberta. Data pooled in three groups of five years, 1969–1983. (All) includes sightings of adult, immature, and falcons of unidentified age groups.

Years	Mean Early Dates			Mid Dates			Mean Late Dates		
	Ad.	Imm.	All	Ad.	Imm.	All	Ad.	Imm.	All
1969–1973	28 Apr	3 May	29 Apr	8 May	15 May	13 May	19 May	25 May	25 May
1974–1978	25 Apr	7 May	24 Apr	7 May	14 May	12 May	19 May	25 May	25 May
1979–1983	28 Apr	4 May	25 Apr	9 May	16 May	11 May	22 May	24 May	25 May

Table 2.2. Age composition of Peregrines in percent of total sightings during spring and fall. Data pooled in three groups of five years, 1969–1983.

Years	Sightings	% Adult	% Immature	% Unidentified
Spring				
1969–1973	135	38	36	26
1974–1978	406	28	47	25
1979–1983	339	29	32	39
Sub Totals	880	32	38	30
Fall				
1969–1973	10	50	30	20
1973–1978	17	6	65	29
1979–1983	24	21	58	21
Sub Totals	51	26	51	23

Table 2.3. Days and hours afield, and Peregrines sighted 15 April– 31 May 1969 –1983.

Year	Days	Hours	Sightings (mean)	Sightings per Hour
1969	22	–	20	–
1970	25	–	15	–
1971	27	–	33	–
1972	25	–	26	–
1973	23	–	41	–
Sub Totals	122	–	135 (27)	–
1974	26	251	46	0.18
1975	29	284	41	0.14
1976	30	307	66	0.21
1977	38	378	163	0.43
1978	34	358	90	0.25
Sub Totals	157	1570	406 (81)	0.26
1979	31	289	57	0.20
1980	34	323	94	0.29
1981	34	308	67	0.22
1982	33	276	58	0.21
1983	35	302	63	0.21
Sub Totals	167	1498	339 (68)	0.23
Totals	446	3068	880 (59)	0.28

Table 2.4. Percent of total field hours for five daily time slots, 15 April–31 May, and sightings of Peregrines per hour. Data pooled for five years, 1979–1983.

0500–0900	8.3	0.28
0900–1200	17.8	0.28
1200–1500	20.5	0.17
1500–1800	27.0	0.20
1800–2200	26.4	0.24
Totals	100	0.23

Figure 2.1. Peregrines sighted during spring migrations in central Alberta. Total sightings include Peregrines of unidentified age class. Data are pooled for five years, 1979–1983.

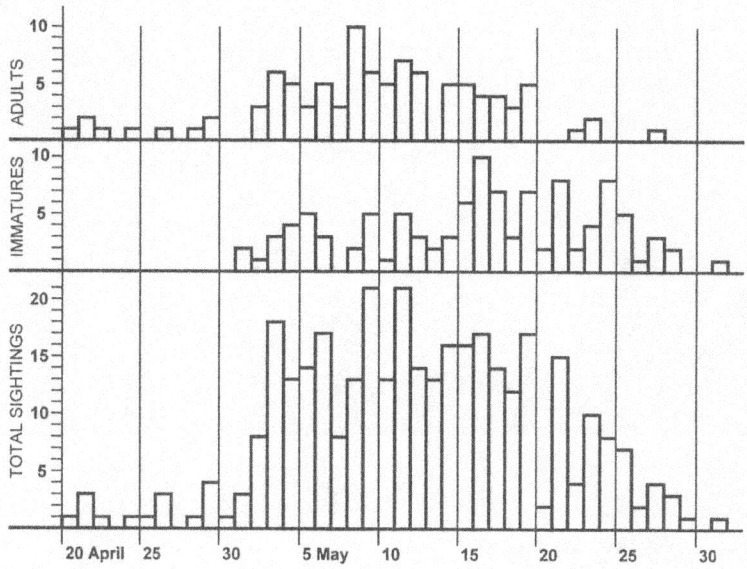

Figure 2.2. Statistical regression line of Peregrine sightings per hour at Beaverhills Lake, Alberta, 1974–1983. The high outlier of 1977 was related to a period of drought with dropping water levels that created vast mudflats at the lake, whereas many area ponds dried up, which led to the concentration of migratory shorebirds and their predators at the lake.

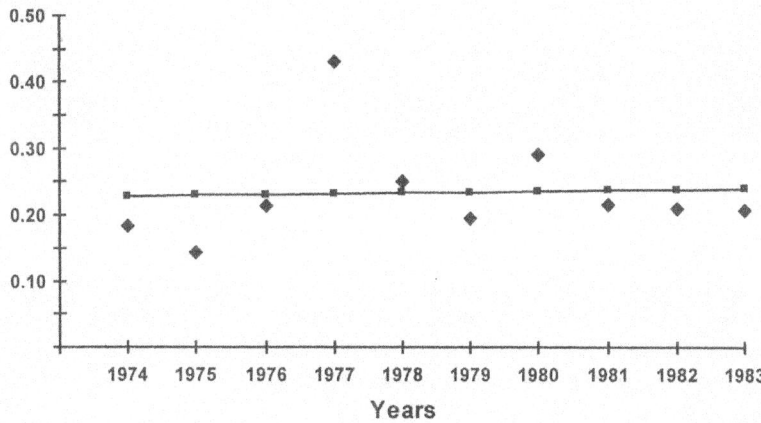

The water levels of Beaverhills Lake, Alberta, are characterized by cyclic highs and lows in response to changes in precipitation and evaporation. Shoreline habitat varies from open mudflats to reedbeds. Hunting Peregrines make use of emergent vegetation to approach by stealth and take their prey by surprise. The most commonly captured shorebird is the Pectoral Sandpiper (Calidris melanotos). (Upper photo: Dick Dekker; lower photo: Brian Genereux).

CHAPTER 3

Hunting strategies of migrating Peregrines

Dekker, D. 1980. Hunting success rates, foraging habits, and prey selection of Peregrine Falcons migrating through central Alberta. Canadian Field-Naturalist 94:371–382.

Abstract – At Beaverhills Lake in central Alberta, during the migration seasons from 1965 to 1979, Peregrine Falcons (*Falco peregrinus*) were observed to attack prey species on 958 occasions. The outcome could be established in 674 hunts, of which 52 ended in the capture of prey, a success rate of 7.7%. Hunting success rates for adult Peregrines and spring immatures were 9.8% and 7.1%, respectively, in 215 and 363 hunts. Fall immatures had a success rate of 2.4% in 42 hunts. A fast stealth approach with the objective of taking the prey by surprise was the basic strategy of the majority (64%) of hunts. Long-distance hunts in which the prey was selected from afar, were significantly more successful than opportunistic surprise attacks. Ducks and small shorebirds made up 94% of prey taken. A few prey were captured on the ground or in shallow water; all others were seized in flight and borne down. The great majority of prey seen to be caught failed in the timely use of escape tactics routinely deployed by their kind. Lone birds were significantly more often killed compared with individuals in flocks.

INTRODUCTION

Although the Peregrine Falcon (*Falco peregrinus*) has been studied very extensively in North America (e.g. Hickey 1969; Cade and Fyfe 1970), no quantitative data on its hunting habits and success rates have been published to date. In particular, how waterfowl and shorebirds are hunted has not been described in detail, although the literature is replete with assumptions and conjecture. A report on the food habits of migrating Peregrines on the Texas Gulf coast was based on the collection of prey remains (Hunt et al. 1975). Primary eyewitness accounts of Peregrines actually capturing prey are few and anecdotal (e.g. Bent 1938; Herbert and Herbert 1965). To my knowledge, the singular exception is Rudebeck (1950,1951), who observed the foraging habits of migrating Peregrines on the south coast of Sweden. He recorded a success rate of 7.5%. However, Fischer (1967) and Treleaven (1961) questioned the value of these data and considered a success rate of 7.5% unrealistically low for a superior raptor such as the Peregrine. Brown and Amadon (1968:75) wrote "probably one attack in four may be successful," but that "much more observation is needed."

Based on a very large sample of empirical data, this paper reports on: (1) The hunting methods of Peregrines preying on waterbirds; (2) The respective success rate of adult and immature Peregrines; and (3) The escape tactics used by ducks and shorebirds attacked by Peregrines. The data were collected from 1965 to 1979 in central Alberta while these falcons were migrating. As reported elsewhere (Dekker 1979), falcon sightings were more common by a factor of 16 to 1 in the spring migration period than during fall.

STUDY AREA AND METHODS

The study area is at Beaverhills Lake in central Alberta, at latitude $53^{0}20$' N. Approximately 10 x 18 km in size, the lake is an eutrophic Ramsar wetland of marshy shallows and cyclic water levels. The open shoreline includes extensive reedbeds, narrow sandy beaches, and rough cattle pastures. Mudflat habitat is ephemeral, dependent upon water level fluctuations and wind direction. The lake's surroundings contain gently rolling agricultural fields with scattered woodlots of aspen (*Populus tremoloides*) and willow (*Salix spp.*). During wet springs, numerous depressions collect snowmelt, attracting ducks and shorebirds that are widely dispersed over the region. In dry years, migrating waterbirds and their predators are concentrated on the lakeshore. For a detailed description of the local avifauna, see the previous chapter and Dekker (1998).

Area visits per year averaged 30 days during the spring migration period from 15 April to 31 May, and 15 days during the fall from 1 September to 15 October. Hours afield varied from three to 17 per day, and included some overnight stays during the spring period. On occasion, the study area was also visited outside the migration periods, but the lake was frozen and snow-covered from early November to late April. Summer sightings of Peregrines were rare and originated from captive-reared birds released in central Alberta by the Canadian Wildlife Service and Alberta Fish and Wildlife Division. In spring and fall, the passage of falcons was uneven and sightings/day varied from none to >20 (Dekker 1979, 1984). To find falcons and maximize chances of observing foraging activity, I used the following methods. Each day I walked 5–25 km, frequently stopping or sitting down to scan the area through binoculars. During rainy, windy or hot weather, I spent many hours sitting in a parked car at points overlooking the lakeshore. Flying falcons were followed in the binoculars for as long as they remained visible. Resting falcons, perched on posts or on the ground, were watched through binoculars or telescope for extended period in the hope of seeing them launch hunting flights that could then be observed from their start, but not necessarily to their conclusion. I also monitored concentrations of shorebirds and ducks for alarm reactions that might give away the approach of predators. Often, a falcon was not discovered until its attack was over, but it sometimes attacked next in an adjacent bay or pond, allowing me to observe the complete hunting sequence from start to finish. Luck, to be in the right place at the right time, was a major factor in this observational study.

In describing and recording Peregrine hunting activity, I used an expanded version of Rudebeck's (1950, 1951) definition of a hunt as consisting of one or more completed attempts by a Peregrine to seize a potential prey. Here, I use the terms hunt and attack as interchangeable synonyms. Aggressive passes at birds that I considered too large to constitute potential prey, such as swans, geese, or cranes, were not included in my tabulations. Male Peregrines are about two-thirds the size of females (Palmer 1988) and they generally take different prey species than the larger females. Male Peregrines often make "playful" passes at medium-sized birds such as American Crows (*Corvus brachyrhynchos*) and large gulls, but I did not consider these as hunts, nor were swoops by either sex at other Peregrines or hawks. A bona fide hunt or attack might include only one swoop at a prey or more than 20 aimed at the same target. A hunt was considered successful if the prey was seized, whether or not the falcon subsequently lost, killed, or ate that prey.

I classified my observations of hunting activity into three different categories. Category A included all hunts of which the outcome was immediately clear because the target was seen to be caught or to escape capture. Category B included hunts in which the target was not in clear view in the concluding stage of the hunt, but the outcome could be deduced from the falcon's subsequent behaviour. If it turned away and selected a different target, the preceding hunt was classified as unsuccessful. On the other hand, if the falcon landed and presently flew off carrying a prey item in its feet, or if it was subsequently seen feeding on a kill, the hunt was termed successful. Category C included hunting flights of which the outcome remained unknown. Hunts in this category could not be used to compute success rates, but they contributed significantly to an understanding of the foraging behaviour of the Peregrine. C hunts included many high-speed, long-distance attacks and stoops from great altitudes, as well as some exceptionally persistent pursuits of prey by one or several falcons in tandem.

On days when little action was seen, observations were written down at the end of the day. On productive days, most observations were recorded immediately after they happened. Some results were compared statistically by Chi-square tests.

RESULTS AND DISCUSSION

Success rates. Peregrine Falcons attacked prey on 958 occasions, but in 284 cases the outcome could not be determined. The remaining 674 hunts resulted in 52 captures, giving a success rate of 7.7%. The respective success rates for categories A ($n = 420$) and B ($n = 254$) were 7.4% and 8.3% (Table 3.1). The difference is not statistically significant ($P > 0.5$). To my knowledge, at the time of this writing, only two other published accounts of hunting by migrating or wintering Peregrines have included hunting success rates. Rudebeck (1950, 1951) computed a rate of 7.5% in 253 hunts by migrating Peregrines in Sweden.

There is no significant difference between his success rates and mine in either category ($P > 0.5$). Clunie (1976) found a slightly higher rate of 9.6% in 62 foraging flights by a single adult female Peregrine wintering in Fiji.

Several researchers (e.g. Berry 1969) have remarked on the apparent lack of skill in hunting by Peregrines during their first few months of independence. This concurs with my limited data for fall immatures; of the 42 hunts seen only one resulted in the capture of a prey. By contrast, during the spring migration period first-year falcons achieved a success rate of 7.1%, close to that of the adults (Table 3.1). The difference was not significant ($P > 0.05$).

Ratcliffe (1962) questioned the intent of some unsuccessful Peregrine hunts that he observed and thought that these falcons were only playing or "practicing." Likewise, many attacks on prey that I saw might be called half-hearted. The explanation is that the hunting activity of a wild Peregrine can be an extended process, lasting several hours with frequent pauses, in particular in habitats where prey are numerous. The initial phase might be considered as a "warming-up" period, until it merges into a relatively brief period of serious hunting. The warming-up phase can be prolonged in immature falcons, and their first attacks are low-level and short-range. Eventually, after several unsuccessful hunts, the falcon sets out on a long-range foraging flight, gaining altitude and looking far ahead for suitable targets. If my observations had been restricted to this final period of serious hunting, the success rates would have been substantially higher than the values recorded here. To the human observer, however, it is not always possible to differentiate between "half-hearted" and serious attacks, even if we go by the only criterion that really counts, whether or not a prey is actually seized. In fact, many unsuccessful attacks were frenzied and spectacular, while some of the kills I watched looked deceptively "half-hearted." Treleaven (1980) introduced the terms high intensity and low intensity hunts. The difference between them is readily recognized by the experienced observer, who has seen many a Peregrine in pursuit of fast-flying prey such as pigeons and high in the sky. But that difference is far less clear if a Peregrine races low along a reedy lakeshore and makes erratic and haphazard passes at shorebirds and ducks flushing just ahead.

It may be logical to conclude that the degree of seriousness of a falcon's attack is a function of the state of its appetite – a hungry falcon is a serious falcon – but this is not necessarily the case for breeding pairs. Nesting falcons are motivated by the need to supply a mate and/or young. As a consequence, they hunt with a greater degree of seriousness than falcons that do not have such duties. This is demonstrated by their success rates. At nest sites on the west coast of Britain, Treleaven (1961) recorded 28 foraging flights by adult Peregrines of which ten resulted in the capture of pigeons, a success rate of 36%. In northern Alberta, I watched parent Peregrines make two single and three cooperative hunts that included two kills. These limited data suggest that the success rate of breeding Peregrines is indeed significantly greater ($P < 0.001$) than that of migrating falcons (Table 3.2). The latter are primarily focussed on covering distance and they forage along the way in an opportunistic manner. If they fail, they move on to try again somewhere else.

In view of the relatively high rate (36%) of hunting success of the breeding Peregrines he watched, Treleaven (1961) expressed doubts about the validity of the 7.5% rate of migrating falcons reported by Rudebeck (1950, 1951). Also Fischer (1967) believed that a success rate of 7.5% was too low, although he did not offer a particular reason. The validity of Rudebeck's data is supported by the near equivalent success rate (7.7%) found for migrating Peregrines in this study.

Hunting success rates may well differ between individual Peregrines, dependent upon their level of skill and experience, and perhaps on specialized hunting techniques. The most dominant falcons take advantage of the highest perches from where they can selectively attack unsuspecting passing prey. A secondary factor in a falcon's kill rate is the vulnerability of prey individuals. For instance, Cade (1960) and Herbert and Herbert (1965) reported on the easy capture of forest birds such as jays crossing wide rivers near Peregrine nesting cliffs. Such prominent hunting perches were not available to the migrating falcons in my study area, where the highest vantage point was a 1.5 m fence post.

Hunting strategies. All observed hunts took place in the same general area, and 95% occurred in April and May. Times of day ranged from 5:45 to 22:00; local temperatures from $-10°$ to $36°$ Celsius. Atmospheric conditions varied from calm and clear to windy with rain or snow, which affected a falcon's mode of flight. During sunny and windy weather, sometimes also in cloudy conditions and even light rain, Peregrines soared and sailed (Dekker 1979). On days with heavy cloud, low temperatures and/or steady precipitation, as well as in early morning and evening, migrating or hunting falcons flew in flapping flight at low or medium elevations.

The vast majority of hunting Peregrines attacked prey species that were on the ground or in shallow water. These hunts could be started from a high soaring position or from active flapping flight, but the falcon's basic strategy was a fast stealth approach, its objective to get as close as possible before being noticed and to take the prey by surprise. The target could be selected from >2 km away, and the final stage of the hunt was a single, direct pass. If the intended target bird fled in time and evaded capture, the falcon might pursue it and make one or more additional passes or swoops, until the prey was either caught or let go.

I divided the Peregrine's hunting habits into several typical methods and strategies (Table 3.3). In the following descriptions, the term "stoop" means that a falcon pulls in its wings, holds them rigid close to the body, and plunges down. The stoop results in a sudden increase in speed, and its trajectory can be either perpendicular or oblique. By contrast, the term "swoop" describes a mode of attack that is essentially different from a stoop because the falcon continues to beat its wings. The swoop also is accompanied by a sudden increase in velocity and a shift in the falcon's direction of flight, but it can be either upward or down. The classic teardrop shaped stoop, with wings pulled in completely, is commonly launched from a high soaring position, but rarely during close pursuit of flying birds. Then, in attempts at seizing the fleeing prey, the falcon swoops at the target. If it fails to connect, it mounts at once, using the impetus of its downward pass to gain altitude again, all the while continuing to beat its wings vigorously. The hunt may end abruptly after a single swoop, or evolve into a long pursuit with the falcon attempting to overtake and seize the dodging prey.

a). Stoop from a soaring position aimed at ground-level targets. This method was used in 22% of hunts in categories A and B and 25% in category C. The altitude of some soaring Peregrines was estimated at >2 km. When over-head, these falcons were just visible in ten-power binoculars. The stoop could be straight or curved, either convex or concave, and the distance covered varied from <10 to >1,000 m. The speed of the stoop was sometimes boosted by bursts of quivering wing beats that involved only the primaries, which were held close to the tail. The majority of stoops levelled out well before the target was reached, and the falcon sailed the last stretch just above land or water until it spread its wings in an abrupt braking action. If the prey flushed just ahead, the falcon attempted to seize it in its feet. If it failed to make contact, it mounted steeply and swooped at the fleeing target. But, usually, the unsuccessful falcon would abort the hunt at once and climb upwind to resume a high soaring position and gain altitude in preparation for its next vertical attack.

b). Stoop from soaring position aimed at flying targets. On 34 occasions high soaring Peregrines stooped at birds flying below at levels ranging from slightly lower than the falcon to far below and just above ground or water. As in the previous group, the stoop could be perpendicular or oblique, straight all the way or curving, and terminate in a single pass or a series of swoops. In some of the most persistent hunts, the swoops became shallower and the attack ended in a level tail chase or a close and erratically twisting attempt at seizing the dodging prey.

c). Long distance flapping flight and descent to ground-level targets. This technique accounted for 18% of hunts in categories A and B and 38% in category C. At the start of the attack, the falcon's altitude was usually moderate, between 50 and 150 m, and the tempo of its wing beats was near normal or even quite slow, until they quickened dramatically, evidently when distant prey was sighted. Well before reaching the target area, the falcon descended, increasing its speed. In rare cases, it stooped down to resume flapping flight just above the ground or water. This secondary stage of the attack often covered distances of >500 m.

d). Low surprise attack aimed ground-level targets. Some of the hunts in this method probably began as long-range flapping flights described above of which the initial higher stage was not seen. However, many falcons, especially immatures, began their foraging by flying at great speed about 1 m high, following the lake shore and flushing prey that happened to rise ahead. The success rate of this opportunistic hunting technique was 4.6%, significantly lower ($P < 0.05$) than the 11.6% and 8.1% of methods *a)* and *c)*, in which the target was selected from afar (Table 3.3).

e). Short-range attack on flying target. After a long-range hunt had failed, or while flapping or sailing at medium or low altitudes, some Peregrines launched opportunistic attacks on prey that happened to fly by or were flushed below. Suddenly changing direction, these falcons flew down in a burst of speed to pursue that prey or to meet it head-on, followed by a quick swoop either up or downward.

f). Take-off from a perch to pursue a flying target. This still-hunting method is commonly used by breeding Peregrines to intercept birds that cross a wide expanse of water within view of the falcon's nesting cliff (Cade 1960; Herbert and Herbert 1965). At Beaverhills Lake, I saw this selective method used seven times. One case involved an American Coot (*Fulica americana*), another a small grebe. The straight unbending flight path of these prey species renders them vulnerable to attack from behind. Both were quickly caught by male Peregrines starting from a fence post.

g). Search for crouching target. After a long-range attack on sandpipers foraging in short grass had failed to flush them out, some falcons engaged in what appeared to be a deliberate search. Flying slowly, head bent down, they suddenly dashed at birds that flushed below, usually evading the falcon.

h). Tenacious pursuit and >10 swoops. About 20% of hunts by fall immatures were exceptionally long chases of shorebirds with repeated swoops. Spring immatures and adult falcons rarely used this method. Five of 11 such hunts became lost to view before I could ascertain their outcome.

i). Pursuit and alternating swoops by two Peregrines. Tandem attacks on the same prey item were seen on 14 occasions. The pairs were of mixed sex and of either age group. During the pursuit, the falcons coordinated and alternated their swoops. While one descended, the other just pulled up. In five prolonged tandem hunts both falcons swooped >20 times each and became lost to view, high in the sky or obscured by reeds or bushes.

j). Other methods. Several Peregrines displayed unusual or unique strategies. **(1)** An immature swooped in a cartwheeling dive with wings spread at swimming phalaropes. **(2)** Another immature hovered over grassland and descended with feet extended in a slow, deliberate manner. Small sandpipers, which had probably crouched at the falcon's approach, got up and fled after it had landed. **(3)** An adult female Peregrine twice left her perch and flew to the lakeshore, where she briefly hovered before returning to her perch. I believe she tried to flush sandpipers she had seen descend. Some sandpipers rose and flew away well before the falcon's arrival. **(4)** A recognizable immature falcon seen at sundown on three consecutive evenings flew low and fast over reedy vegetation, swooping in a peculiarly steep rise-and-plunge fashion at Red-winged Blackbirds (*Agelaius phoeniceus*). **(5)** On days when conditions were not suitable for soaring, several Peregrines gained height by flying in circles or in a zigzagging pattern, prior to launching long distance attacks on shorebirds and ducks.

Prey selection. The targeted prey species could be identified in 641 hunts: 20% were ducks; 56% small shorebirds; 9% open-country passerines; 3% gulls; and

2% miscellaneous species (Table 3.4). An additional 10% were either ducks or shorebirds. Together these two groups made up 86% of all prey hunted. Generally, female falcons captured ducks, while males took shorebirds and passerines. Some Peregrines alternately attacked ducks or shorebirds, evidently selecting their target in an opportunistic manner, others pursued the same kind of prey in >10 consecutive hunts until a kill was made. A few falcons appeared to have specialized on gulls.

a). Ducks. All adult Peregrines seen during early spring and most adult and immature females observed later in the season preyed almost exclusively on ducks up to the size of Northern Pintail (*Anas acuta*). Mallards (*Anas platyrhynchos*) were rarely taken and only the hens. The hunting methods varied greatly (Table 3.3). Of 131 hunts aimed at waterfowl, 40% were long-distance flapping flights or low-level attacks on resting or feeding flocks along the lakeshore or in area wetlands. Alarmed ducks flushed and splashed down into the nearest water. If approached directly by a low flying falcon, swimming ducks dived, kicking up water. From very shallow surfaces, the ducks flushed and instantly splashed down again if the falcon tried to grab them. The common escape tactic employed by flying ducks was to drop like a falling stone at the last moment, either into water or vegetation. Some flying ducks, especially if attacked from below, dodged the falcon by veering aside. To evade a falcon that stooped at them from a superior altitude, a group of high-flying pintails mounted steeply. Pursued further, they dove down at speeds matching that of the falcon. Several immature Peregrines attacked flying ducks after sundown and at dusk (Dekker 1979). The following descriptions of specific incidents are examples of the varied methods falcons use in hunting waterfowl.

1). 4 May 1975, 11:00. An adult male Peregrine took off from a fence post to intercept a coot that flew low between two bodies of water. The coot did not dodge and was seized directly from behind and borne down.

2). 10 May 1976, 11:00. An immature female flew low and fast over a pasture toward the lake. A pintail drake, flushing just ahead of the falcon, was seized directly and borne down onto the muddy shore, where it was killed in a brief wing-flapping struggle.

3). 15 May 1976, 18:00. A high-soaring Peregrine travelled >1 km in a shallow stoop, assisted by bursts of vibrating wing beats, to meet a dozen ducks head-on, aiming its attack at a bird on the outside of the string. The duck dropped perpendicularly, but was overtaken by the stooping falcon, which seized the dodging prey after a twisting aerial pursuit, and carried it far inland, descending in a long glide.

4). 22 May 1976, 19:00. An immature falcon, flying at an estimated 200 m high in a slow, almost hovering manner, suddenly descended steeply to flush half a dozen pintails from a stubble field. One of the ducks was seized about 3 m above the ground and borne down.

5). 18 May 1977, 20:00. Two immature females, following each other about 50 m apart, were flying over agricultural fields, when they abruptly turned direction to meet a pair of Northern Shovelers (*Anas clypeata*) head-on. The lead falcon seized the duck but released it again when the second falcon swooped it at. The first one recaptured the duck after a brief chase and bore it down steeply to a ploughed field. Both falcons fed on the kill.

6). 30 April 1978, 13:00. An adult female flying over the still partly frozen lake stooped at a pair of Lesser Scaup (*Aythia affinis*) that passed below the falcon, flying into the opposite direction. The ducks splashed down in open shore water. When the falcon made a pass at one of the ducks, the other one got up and flew off, to be pursued by the falcon. Just before it was overtaken, the duck dropped into a pool of water, evidently not deep enough to submerge. The falcon, its wings extended, landed on the prey and presently dragged it onto the ice.

7). 17 May 1978, 11:00. An adult female, soaring over the lakeshore, flew down obliquely in pursuit of a pair of Green-winged Teal (*Anas crecca*) passing low over reedbeds. The drake failed to dodge, was seized directly from behind and carried to shore.

8). On eight occasions during May 1980, (after the collection of data for this paper had been terminated), I observed an adult female start from a fence post to meet ducks crossing 1 km of open pasture between the lake and an inland pond. The falcon approached the ducks head-on and swooped steeply upward 15–30 m, attempting to seize its target from below. Most ducks dodged the falcon, veered aside and were let go. Two failed to evade the upward swoop. A drake Lesser Scaup was seized from behind; a female pintail from the side. None of the eight ducks attacked reverted to their usual escape tactic of plunging down, probably because the attack took place over open grassland and the falcon was flying lower than the ducks.

b). Shorebirds. Migrating shorebirds varying in size from Black-bellied Plover (*Pluvialis squatarola*) to Least Sandpiper (*Calidris minutilla*) were hunted by Peregrines of both age groups and sexes. The Pectoral Sandpiper (*Calidris melanotos*) was the single most preyed-upon species, constituting 53% of 32 identifiable shorebird captured. Most hunts (>90%) were directed at feeding or resting flocks, and surprise was the principal strategy in 26 of 36 successful hunts observed. While the Least and Pectoral Sandpipers were seen to crouch down to avoid detection, other shorebird species, uttering sharp calls, took to the air and drew together in defensive formations that coursed back and forth until the threat had passed. Climbing high into the sky, these compact flocks were seldom attacked. The alternating light and dark flash pattern of wheeling shorebirds may well have an additional function as a long-distance danger signal alerting other flocks that rise in response. After Peregrines have hunted aggressively over a particular section of the lake, the local shorebirds become agitated and often leave for other feeding grounds.

Sandpipers that flushed well ahead of an approaching falcon veered sharply out of harm's way and gained altitude quickly. Those that flushed late routinely avoided capture by dropping onto the ground or into shallow water, taking off again instantly into the opposite direction, while the Peregrine overshot its mark. Most falcons abandoned the attack at once and flew on. If the sandpiper had not risen in time but stayed down, the falcon mounted steeply and doubled back, attempting to grab the crouching prey, or to pursue it in case it tried to escape. It is possible that some sandpipers were raked by the falcon's claws and unable to rise. Others were prevented from rising because the falcon was directly overhead. Sandpipers that had dropped into water to escape the initial attack lurched about erratically to dodge the falcon's repeated passes. In one such hunt observed at close range, the sandpiper managed to find safety in a tussock of rushes. In another incident, the splashing and diving sandpiper was eventually seized by one wing and carried aloft.

After prolonged chases, some sandpipers descended steeply and took cover in vegetation. The Peregrine might then circle the spot and swoop repeatedly in a vain effort to flush out the hiding target. Other shorebirds successfully evaded 10–20 swoops until the falcon gave up. A Short-billed Dowitcher (*Limnodromus griseus*) outclimbed a pair of cooperating immature Peregrines in an aerial chase that started at water level and reached an altitude of >150 m. The outcome of five other prolonged chases of shorebirds by single falcons or pairs could not be ascertained as the birds became lost to my view. Once they had attained the advantage of superior height, Peregrines could quickly overtake and close in on a fleeing shorebird, but they lacked manoeuvrability. Between swoops, pursuer and pursued were often >50 m apart. Of 36 shorebirds caught, 27 were sandpipers. None of these 27 hunts involved a long pursuit, and all of the 14 sandpipers caught under perfect viewing conditions were seized at the falcon's first pass. In a sense, these sandpipers failed to deploy the usual escape tactics of their species, namely a well-timed plunge into water or cover, followed by a fast and erratic getaway.

The following eyewitness anecdotes include typical and not so typical examples of the methods Peregrines use in hunting shorebirds.

1). 8 May 1972, 16:30. A high-soaring adult female stooped low over a wetland at a group of Pintail ducks that splashed about but did not rise. As the falcon mounted steeply again, a single Pectoral Sandpiper flushed under her. She stooped and caught it just above the water.

2). 16 May 1974, 16:30. An immature female falcon, flying at a moderate altitude, descended gradually to the lakeshore >1 km away. A lone Black-bellied Plover flushed ahead of the falcon and was seized directly from behind 3 m above the ground.

3). 13 May 1975, 14:00. An adult male Peregrine raced low over a marsh where dowitchers flushed. For a brief moment, the falcon was out of view behind some tall reeds, it then reappeared and abruptly halted to hover at 2 m above the water. Staying suspended for several seconds, it suddenly plunged in, feet first, submerged up to its belly. When it rose again, the falcon had a dowitcher in its feet. Apparently, the dowitcher had splashed down and dived into the water to evade the attacking falcon.

4). 20 May 1975, 17:45. An adult female Peregrine, soaring high over inland fields, stooped at a shallow angle and sailed at great speed for a horizontal distance estimated to be 1.5 km. The last 200 m of its trajectory were very low over the lake. Half a dozen Black-bellied Plovers flushed in a panic from the muddy shore of an open point. One of them was seized at once.

5). 13 May 1977, 16:15. An immature female flew at about 30 m over a pasture, flushing hundreds of sandpipers that had been feeding on wind-grounded lake flies (a large species of chironomid midge). She descended to overtake a Pectoral Sandpiper flying low over the grass, seizing it directly from behind.

6). 23 May 1977, 21:45. Just after sundown, an immature falcon took off from a fence post and flew about 600 m, climbing steadily to a height of about 150 m, heading for a dense globular flock of about 100 Buff-breasted Sandpipers (*Tnyngites subruficollis*) that had risen in alarm at the falcon's approach. One of the sandpipers was seized directly from the middle of the flock, which scattered upon impact.

7). 9 May 1978, 11:30. An immature male and an adult male Peregrine were perched on adjacent fence posts. The male flew to the lakeshore and attacked a Lesser Yellowlegs (*Tringa flavipes*) that flushed ahead. The female falcon joined the chase and each Peregrine made three or four swoops until the male seized the prey. He released it again when harassed by the female. Pursued by both Peregrines, the yellowlegs flew on for a short distance and was captured by the female at her next pass.

8). 15 May 1978, 13:00. A soaring immature female falcon stooped and raced low over a cattle pasture from where hundreds of sandpipers flushed. One of them dropped into the grass a few metres ahead of the pursuing falcon, which doubled back to seize the prey on the ground.

During May 1980, after data collection for this paper had been finished, I saw an additional 17 shorebird hunts of which 13 involved flocks of Pectoral Sandpipers foraging on grounded midges clinging to the grass (Dekker 1985, 1998). Five of these attacks ended in captures. The falcons used a variety of hunting methods but in all cases their final approach was very low over the grass, either in a fast glide or flapping flight. In one successful hunt, a fleeing sandpiper was seized in the air. In the other four hunts, the sandpipers dropped into the grass just before being overtaken. The falcon then swooped 3–30 m upward over the spot and quickly dropped down again to grab the prey on the ground. The easy success of the falcons in this series had to do with several environmental factors. (1) The pasture grass was rather long, with tussocks of 10–50 cm, which helped Peregrines to conceal their approach. (2) The dense grass may have given the sandpipers a false sense of refuge and safety, which might explain why they dropped down to crouch and why they did not get up again immediately after the falcon had overshot its mark. (3) And most importantly, due to the exceptionally rich feeding opportunities in the form of a massive hatch of midges, the sandpipers had probably built up a heavy fat load, which can be expected to slow down their take-off speed and reduce their manoeuvrability.

c). Gulls. Several species of gull are common in the study area, but they were not often hunted by Peregrines. Gulls dodged nimbly when attacked from above, evading a falcon's stoop by mounting steeply. A Ring-billed Gull (*Larus delawarensis*) even avoided capture by a closely pursuing Peregrine that attempted to strike it from below. An immature female Peregrine consistently hunted Franklin's Gulls (*Larus pipixcan*) by meeting them head-on or from the side, and rising up to them from below. On two occasions, a high-soaring Peregrine stooped at resting gulls without making a capture. Two fall immatures chased gulls with great persistence and over long distances, both disappearing from my view before I could ascertain the outcome. Rose (1965) followed a similar hunt by car and recorded the capture of a Franklin's Gull by a fall immature Peregrine after a chase of 6 km.

The only successful gull hunt I saw in this study took place at 11:30 on 20 May 1978. An immature falcon that had already made several fruitless rushes at flocks of resting gulls, took off from its perch on a stone to climb steeply out over the lakeshore against the wind. Numerous Franklin's Gulls were hovering and riding the wind, hawking for flying insects. One of them was seized directly from behind. Carrying the limp prey, the Peregrine turned back to land. It was obvious that the gull, facing into the breeze, had failed to see the Peregrine coming up from behind and below.

d). Small passerines. Although blackbirds were attacked by Peregrines of both sexes, smaller open-country passerines such as sparrows, longspurs, and larks were exclusively hunted by male falcons. The success rate of 55 passerine hunts was only 3.6%, lowest of the four different groups of prey species (Table 3.4). During the cold spring of 1979, when shorebirds were scarce, there were very large flocks of Snow Buntings (*Plectrophenax nivalis*), containing several thousands of birds, on area fields. They were hunted by a team of adult male Peregrines. Launching their stoops from a high soaring position, they flushed the flocks from the ground and swooped in tandem at individually fleeing buntings. Once air-borne, they dodged effectively. After repeated attacks over a period of one hour, the flocks became extremely agitated. Their feeding routine was disrupted by continuous alarms, real or false. The first birds to land would rise again at once, taking the flock with them before the last in the group had a chance to touch down. On two occasions, the falcons stooped at dense flocks in flight, splitting them and pursuing single birds, again without success. The only successful hunt I recorded occurred at 12:00 on 5 May 1979, when a soaring adult male Peregrine stooped at a feeding flock of buntings and seized one of them the moment the flock got off the ground.

To escape detection by hunting falcons, small passerines including Snow Buntings commonly tried to hide. After this study was terminated, I saw an adult male Peregrine make a deep stoop at a compact group of buntings crouching together on a 4 m high berg of ice floes stranded on the lakeshore. The buntings did not flush, and the Peregrine plucked its prey from the ice while the other buntings remained immobile. It is tempting to speculate that the whiteness of the ice had given the (white) buntings a heightened, but false, sense of camouflage and safety.

e). Crows, hawks, and owls. Although Peregrines swooped at crows on some 20 different occasions, only four interactions qualified as bona-fide attempts at capture. An adult female stooped from a soaring position at a lone crow flying at a height of circa 300 m over the lake. After 10–12 failed swoops, the falcon ended up tail chasing the target and trying to grab it from the side or below. The crow was let go after it got close to an inland woodlot. Three other crows were vigorously attacked by immature falcons. Two of them found refuge in trees after dodging 8–10 swoop during chases covering 1–2 km. The third crow managed to escape despite being handicapped by very ragged pinions. In open country, the approach of falcons induced flocks of crows to climb high into the sky or take cover in trees, on fence lines or buildings.

Peregrines often swooped or stooped at other raptors, but I saw only four attacks that were directed at species small enough to be considered potential prey. An immature female Peregrine, soaring very high, stooped at and missed a Sharp-shinned Hawk *(Accipiter striatus)* soaring below the falcon. On two other occasions, soaring immature Peregrines attacked unidentified small raptors high in the sky, driving them down in aggressive pursuits and swooping 8–12 times, but desisting close to the ground.

A female Peregrine pursued a Merlin *(Falco columbarius)* over open grassland, making three shallow passes, each time falling several metres behind, but gaining again slowly. Eventually, about 1 km away from my observation point, the Merlin climbed at a steep angle, reaching an altitude of 200–300 m, until it was almost overtaken by the closely following Peregrine. Hard-pressed, the Merlin dropped down perpendicularly with the larger falcon in pursuit and both disappeared from view behind a rise in the land. Presently, I meant to see the Merlin flying away free, but I could not be sure about the final outcome of this deadly race.

Short-eared Owls *(Asio flammeus)* were the subjects of three unsuccessful attacks by immature Peregrines. In two of these hunts, the falcons swooped two or three times, but missed their mark. The third owl was attacked by a pair of falcons working in tandem. Climbing steadily and dodging its attackers in what seemed a very relaxed manner, the buoyant owl evaded 18–22 alternating stoops until the pair gave up.

f). Miscellaneous. The following observations of Peregrine food habits were not included in the tabulations for this paper. **(1)** Two immature female falcons, observed through the telescope, were walking on short grass that was covered with swarming midges *(chironomids)*. The falcons appeared to be picking up insects in their bills. **(2)** On a warm day in early fall, an immature falcon tried unsuccessfully to capture dragonflies. **(3)** A summering male Peregrine, probably from captive-reared stock, caught three dragonflies in quick succession and ate them on the wing. **(4)** An adult female Peregrine and a recently released immature male from captive stock flew down from fence posts to seize small rodents, probably voles. **(5)** On two occasions, I saw immature falcons land on and start feeding on the carcasses of ducks that had died during an outbreak of botulism on the lake.

Killing practices, food utilization, and piracy. Several authors (e.g. Godfrey 1966; Beebe 1974) claim that Peregrines strike their prey in the air, sending them tumbling to earth, either dead or badly injured. By contrast, in all successful aerial captures in which I could clearly see the moment of impact, the prey was simply seized in the falcon's feet and carried to earth. I saw no evidence that they were knocked down, although the plunging escape routines used by waterfowl and shorebirds might look as if they had been struck. However, aerial hits can and do happen. Of 18 kills observed by Rudebeck (1950, 1951), two involved birds that fell to earth lifeless. While observing wintering Peregrines in the Netherlands, I saw an adult falcon strike a Rock Dove (*Columba livia*), resulting in a burst of feathers, although the pigeon managed to reach the cover of woods. There, I found it dead minutes later. No doubt some Peregrines kill or maim their prey in the air by raking it with their claws, but such hits may be far less common than is generally believed. The astute observer of hunting Peregrines can only conclude that an aerial blow was struck if feathers were seen to fly, or if the prey is subsequently found to have been crippled or killed.

Well after this study had been terminated, I had a very close view of an adult Peregrine stooping at a flushing flock of Semipalmated Sandpipers (*Calidris pusilla*) on the shore of Beaverhills Lake. One bird fell back into the water. The falcon retrieved the apparently lifeless carcass, landed briefly on a nearby rock and presently flew off again, resuming its migration and leaving the dead sandpiper behind.

Some prey that I saw seized by Peregrines appeared to be instantly stunned. Two falcons that caught sandpipers in the air, holding them in their feet, brought their prey forward and bent down to bite it. However, two Pectoral Sandpipers flapped one or both wings while being carried to the ground. Three other sandpipers escaped after the falcon had landed with its prey. Two were recaptured after a short chase, a third managed to find refuge in bushes 25 m away, where it stayed put despite the falcon's repeated passes at its hiding spot. Two ducks and one Lesser Yellowlegs seized in the air by falcons were held fast for a moment and released again. Flying away as if unharmed, they were quickly recaptured. However, of two other ducks that were grabbed in the air and released, one fell into reeds as if dead, the other flew away and escaped by dropping into water at the next swoop of the pursuing falcon. A small grebe lost a puff of feathers when struck from behind by a falcon. Continuing its flight, the grebe dropped into bushes just before being overtaken again.

Ducks and coots that were brought down to the ground were despatched after a brief wing-flapping struggle, but they were not always eaten. An adult male Peregrine that killed a coot abandoned it after a few minutes. On 8 May 1980, I observed an adult female falcon seize a drake Lesser Scaup and kill it on the ground. Presently, the falcon flew to a nearby fence post, and after some time began attacking other ducks. It is possible that these Peregrines would later return to their prey. At the time they made the kill, they might still have been in a "warming-up" phase and not hungry enough to start eating.

Birds up to the size of Pectoral Sandpipers were always consumed completely except for the feathers, some entrails, one or both legs, and the mandibles. Of ducks, gulls, and the larger plovers only the neck and breast meat were consumed at the first meal, but the falcons might return to their kills two or three hours later.

Brown and Amadon (1968) gave the daily food requirements for captive Peregrines as 104 grams of meat per day during winter. They further suggested that a substantial kill might last large falcons in the wild for several days. However, the migrating Peregrines in this study appeared to kill two or more birds daily. Duck-hunting falcons were seen to eat in early morning, late morning, and again in late afternoon or evening. When my presence near an early morning kill made it unavailable for a female falcon's second meal, she caught another duck before leaving on migration. Most probably she hunted again at the end of a long day of travel. Shorebird hunters made several foraging flights per day, perhaps three or four. Immature falcons that had consumed an entire Stilt Sandpiper (*Micropalama himantopus*) in one instance and most of a Black-bellied Plover in another began hunting again after a rest of two hours. Three falcons that I watched eating a sandpiper or small passerine chased prey species immediately afterwards.

The waste factor in falcon predation on large prey species can be high, especially during migration times. An immature female falcon ate only the skin of the neck and part of the breast meat of a drake pintail before she left the study area. The remains of Peregrine kills are utilized by crows, Black-billed Magpies (*Pica pica*), buteo hawks, and Northern Harriers (*Circus cyaneus*), which wait near feeding Peregrines and move in soon after the falcon flies off. Six harriers surrounded a female falcon plucking a duck. When an American Rough-legged Hawk (*Buteo lagopus*) hovered low over a feeding female Peregrine, she mantled over her prey until the hawk flew on. Female Peregrines are quite capable of defending their kills against the piracy attempts of Red-tailed Hawks (*Buteo jamaicensis*), Swainson's Hawks (*Buteo swainsoni*), and Rough-legged Hawks, as well as harriers, but male Peregrines are not. An immature male that captured two teal in five minutes lost both at once to buteos.

Even when male Peregrines capture small shorebirds, the scavengers are quick to approach. It results in the hurried departure of the falcon, which is forced to transport its prey over long distances before settling down to feed. Intraspecific food theft is also common. The females attempt to rob the males and nearly always succeed. Twice I saw an adult female rob immature males of their just-caught prey. However, an immature female vainly pursued a prey-carrying adult male for >2 km, steadily climbing high into the sky. Another adult male ended up dropping his sandpiper at the vigorous prodding of an immature male, but the prey fell into reeds and was not retrieved. To prevent piracy attempts by buteo hawks as well as conspecifics, male Peregrines commonly take their prey high into the sky and pluck it while soaring or sailing (Dekker 1979).

Peregrines routinely rob other raptors. An adult male chased a Sharp-shinned Hawk until it dropped its prey that was caught in mid-air by the falcon. On two occasions, Peregrines pursued Merlins forcing them to jettison freshly caught sandpipers. Food carrying adult male harriers appear less quick to surrender. Two harriers managed to hold onto their food items despite repeated violent swoops by Peregrines. However, a third harrier gave in and dropped its load, which fell into reeds and was lost to both raptors.

Selective effect. Rudebeck (1950-1951) and Eutermoser (1961) have presented evidence of a selective effect of Peregrine predation as shown by a relatively high percentage of abnormal individuals among the prey they saw taken by Peregrines. I could not detect any physical deformities in birds seized by Peregrines in this study, except for one duck that was crippled by botulism poisoning. However, the majority (81%) of birds captured failed to use escape tactics, such as dodging or plunging down, routinely deployed by their kind. These timely tactics were typical of the great majority (94%) of 420 birds attacked under category A conditions that allowed me to see exactly what happened. Only six (1.5%) of these birds were caught. By contrast, all of the 26 birds that failed to dodge in the correct way were seized at the falcons' first try. The difference is highly significant ($P < 0.001$). These 26 victims were fleeing on a straight course. Some of them dodged by dropping into the grass, but they did so too soon and they failed to get up and away the moment the falcon overshot its mark. (For specific examples, see the descriptions of successful hunts in the previous sections). The exact reason for their faulty responses is not known. I hypothesize that these prey individuals were less fit than they should have been, perhaps because they had gorged on the abundance of grounded midges. (High body mass and excess fat loads have since been identified as critical factors negatively affecting a shorebird's manoeuvrability and hence its predation risk (Dietz et al. 2007)).

It is possible that some birds seized directly from behind did not see the Peregrine coming. This was almost certainly the case of the gull taken from behind and below. Other prey, such as the plovers and the Snow Bunting, were surprised at such close quarters that there was no chance to dodge. The majority of prey, however, appeared to have had fair warning of the approaching danger and enough time to dodge; yet they did not do so or were too late to try. The conclusion seems justified that most of these birds showed abnormal escape behaviour. I believe that no prey species is capable of flying faster than a Peregrine, and that birds under attack can avoid capture only by taking evasive action or seeking cover. In this study, targets that dodged were rarely caught and quickly left alone. However, there are circumstances in which Peregrines are quick to take advantage of birds that happen to be in a vulnerable position, such as jays and thrushes migrating over water. As reported by Herbert and Herbert (1965), Peregrines catch these out-of-habitat species despite their attempts at dodging.

Birds that flush well ahead of an approaching Peregrine have a better chance of reaching their top speed and discouraging pursuers than birds that are surprised at close range. Therefore, early warning is of critical importance. Powell (1974) proved experimentally that feeding flocks of Starlings (*Sturnus vulgaris*) responded more quickly than single birds to a flying model hawk, and that individuals in a flock spent less time in surveillance and more time feeding.

Assuming that the same applies to other flocking species, lone ducks or shorebirds should be more vulnerable than flocks. Out of 226 surprise attacks by Peregrines, observed under category A conditions, and directed at feeding or resting prey species, 17 involved lone individuals and 209 were aimed at flocks. The respective numbers of kills were four (24%) and 13 (6%), suggesting that lone individuals are significantly ($P < 0.02$) more at risk than individuals in flocks.

It is improbable that Peregrine predation has any selective effect if the falcons restrict their attacks to prey that happens to be in an out-of-habitat and therefore vulnerable position, such as forest birds migrating over water. Selection is more likely to occur if falcons attack prey species in their natural habitat of preference. For that reason, this study was well suited to examine the selective effect of Peregrine predation, especially since many falcons were believed to be in their "warming-up" phase and to hunt in a "half-hearted" manner. I hypothesize that these Peregrines foraged more selectively than a falcon that is deadly serious. Furthermore, as suggested by the results of this study, migrating falcons appeared to "test" their prey by a single stoop or pass. Targets that dodged in the normal manner were soon left alone. After a number of such half-hearted attacks had failed, the falcons in this study launched long-range stealth hunts and became more tenacious in pursuit. These deadly serious Peregrines proved capable of catching some prey before they had a chance to dodge or despite their dodging.

Table 3.1. Number of hunts and success rates of Peregrine Falcons migrating through central Alberta. The outcome of category A hunts could be determined at once; the outcome of hunts in category B was not immediately apparent but became known after follow-up investigations. See Methods section.

	Number of Hunts			Success Rate (%)		
	A	B	A+B	A	B	A+B
Age group						
Adults	126	89	215	8.7	11.2	9.8
Spring Immatures	245	118	363	7.3	6.8	7.1
Fall Immatures	25	17	42	4.0	–	2.4
Unidentified	24	30	54	4.2	10.0	7.4
Totals	420	254	674	7.4	8.3	7.7

Table 3.2. Comparison of hunting success rates of adult Peregrines on migrating or on wintering grounds with the success rate of breeding adults. The difference between breeding and non-breeding falcons is highly significant ($P < 0.001$).

	Hunts	Success %
Adults during migration in central Alberta (This study)	215	9.8
Single adult wintering in Fiji (Clunie 1976)	62	9.6
Breeding adults in Great Britain (Treleaven 1961)	28	35.8
Breeding adults in Northern Alberta (Dekker, Unpublished)	5	40.0

Table 3.3. Hunting methods, number of hunts, and their success rates. (Successful hunts are in brackets.)

Hunting Techniques	Hunts (Kills)	% Success
Stoop from soaring position at ground level target	112 (13)	11.6
Stoop from soaring position at flying target	34 (3)	8.8
Long-distance flapping flight and descent to ground level	124 (10)	8.1
Low surprise attack aimed at ground level target	197 (9)	4.6
Short-range attack on flying target	151 (10)	6.6
Take-off from perch to pursue flying target	7 (2)	28.6
Unknown approach	13 (2)	15.4
Other methods	36 (3)	7.9
Totals	674 (52)	7.7

Table 3.4. Number of hunts and success rates of Peregrines attacking various prey species. Sixty-one hunts that were directed at either ducks or shorebirds were divided proportionally between these two groups. See also Table 3.2.

	Hunts	Kills	Success %
Ducks	153	13	8.5
Shorebirds	400	36	9.0
Gulls	21	1	4.8
Small passerines	55	2	3.6
Other prey	12	0	–
Species unknown	33	0	–
Totals	674	52	7.7

In 1987, Beaverhills Lake was designated a Ramsar Wetland of International Importance, and in 1996 it became a Western Hemisphere Shorebird Reserve. Unfortunately, by September 2006, its once 140 square kilometre expanse had dried up completely, due to upstream water withdrawals superimposed on a series of years with less than average precipitation. In the spring of 2007, a huge snowmelt brought the lake back up to about one third of its former size, but by late summer of 2008, it had again shrunk to the zero point. In the past, the shallows attracted huge flocks of migrating waders, such as these Long-billed Dowitchers (Limnodromus scalopaceus). *(Photo: Brian Genereux).*

CHAPTER 4

Duck hunting by migrating and wintering Peregrines

Dekker, D. 1987. Peregrine Falcon predation on ducks in Alberta and British Columbia. Journal of Wildlife Management 5:156–159.

Abstract – Peregrine Falcons (*Falco peregrinus*) migrating in Alberta attacked ducks 275 times and captured 25. Peregrines wintering in British Columbia caught 10 ducks in 43 hunts. Foraging methods differed in the two areas. The wintering falcons mainly used still-hunting techniques and launched surprise attack on ducks from high trees. Their success rate was significantly higher than the overall rate of migrating falcons, which mainly hunted by long-distance search flight. Eighteen ducks were seized in the air and borne down, one was struck in the air and fell to the ground, 16 ducks were seized on or near the ground or in water. Individual Peregrines caught up to three ducks/day, but scavengers consumed most of the prey. On the British Columbia coast, six falcons lost just-caught ducks to Bald Eagles (*Haliaeetus leucocephalus*) and Golden Eagles (*Aquila chrysaetos*). In Alberta, Peregrines were often robbed by buteo hawks (*Buteo* spp.). The hypothesis is advanced that the loss of prey to larger raptors has contributed to the Peregrine's use of hunting habits that reduce that risk.

INTRODUCTION

The foraging habits of wild Peregrine Falcons preying on a wide variety of land birds have been documented for many regions of the species' nearly worldwide range (Rudebeck 1951; Herbert and Herbert 1965; Monneret 1973; Clunie 1976; Treleaven 1980; Bird and Aubrey 1982; Thiollay 1982; Vasina and Straneck 1984). Hunting strategies on waterfowl have been reported only from Alberta (Dekker 1980). This paper analyzes and compares duck hunting techniques, predation rates, and prey utilization of Peregrines migrating through central Alberta and Peregrines wintering in coastal British Columbia. The study was partially funded by the Alberta Fish and Wildlife Division, the World Wildlife Fund Canada, and the Alberta Recreation, Parks, and Wildlife Foundation. R. W. Nelson and W. D. Wishart reviewed the manuscript.

STUDY AREAS AND METHODS

The Alberta study area was 60 km east of Edmonton and consisted of roughly 10 x 2 km of grain fields and open pastures bordering on Beaverhills Lake, that attracted numerous waterfowl and shorebirds during migration periods. Peregrine Falcons, adults and immatures, occurred from mid April to late May, and from early September to early October (Dekker 1979, 1984). Resting falcons, sitting on posts or on the ground, were often watched until they left of their own accord. Flying falcons were followed through binoculars for as long as they remained visible. From 1969 to 1986, the study area was visited on 529 spring days (15 April–31 May) and on 249 fall days (1 September–15 October).

The British Columbia study area was roughly 3 x 1 km of gently sloping farmlands surrounded by wooded hills near Victoria, Vancouver Island. During winter, low-lying fields were flooded by rain, attracting hundreds of ducks. One or more adult Peregrines (*F. p. pealei*) wintered in the area and perched in trees overlooking the wetlands. The falcons were observed on a total of 23 days during January and February 1980–1986.

A hunt is defined as a completed Peregrine attack on a duck or a group of ducks, involving one or more swoops on the same target. Hunts that resulted in the capture of prey were termed successful although some ducks escaped or were released alive after capture.

Data were compared statistically by independence tests using G-statistics.

RESULTS AND DISCUSSION

Hunting techniques. Sixteen of 25 ducks captured in Alberta were seized in the air and borne down; nine were seized on or near the ground or in shallow water (Table 4.1). If approached or pursued by Peregrines, flying ducks attempted to reach water and usually splashed down before being overtaken. Ducks that had flown up from a body of water often returned to it after a falcon approached them. Swimming ducks dived when attacked. However, three ducks were seized and killed after they had dropped into water too shallow to submerge. Three ducks flying over water or reeds were seized but released again because the Peregrines, losing altitude, were unable to carry their prey to land. A drake Northern Pintail (*Anas acuta*), seized and borne down over a flooded field, struggled free and submerged or hid in weedy vegetation.

Migrating Peregrines in Alberta began 94% of hunts while they were in flight – either soaring, sailing, or flapping their wings – at altitudes ranging from just above ground to >1.5 km high. They often attacked flying ducks head-on and approached at a lower level than the ducks, probably to prevent their plunging escape attempts. Examples of successful duck hunts using various techniques were given in Dekker (1980).

In British Columbia, Peregrines seized seven ducks on the ground or at the moment they flushed off the ground, and three in the air after low pursuits (Table 4.1). In unsuccessful hunts, ducks routinely dodged a falcon's swoop by plunging into water or reeds, and the split-second timing of this evasive tactic made it seem as if the duck was struck down, but only one aerial knock-down was observed; a Green-winged Teal (*Anas crecca*) fluttered to the ground after it had been crippled by the falcon in a head-on, upward swoop. Aerial knock-downs were also rare for other prey species captured by Peregrines (Dekker 1980).

All duck hunts by these coastal falcons were launched from high trees after the prey had apparently been spotted by the falcon from its perch. Similar "still-hunting" techniques have been reported for breeding Peregrines that commonly initiate attacks from high nesting cliffs on vulnerable forest birds flying over water (Herbert and Herbert 1965). Hunting success rates of breeding Peregrines range between 25% and 52% (Treleaven 1980; Bird and Aubrey 1982; Thiollay 1982), significantly higher ($P < 0.001$) than the 7–10% success rates of migrating Peregrines (Rudebeck 1951; Dekker 1980). In this study, the success rate of the still-hunting Peregrines wintering in British Columbia was significantly higher ($P < 0.01$) than the kill rate of the migrating falcons in Alberta, which used active, long-distance search methods. However, the few Alberta falcons that did use "still-hunting" techniques achieved similar kill rates as the coastal falcons. The highest perches in the Alberta study area were 1.5 m fence posts. Five out of 17 attacks on ducks that were started from fence posts were successful, a rate of 29%, significantly higher ($P < 0.01$) than the overall rate of 9%.

Prey utilization and losses to scavengers. In addition to the 35 ducks captured by Peregrines, ten ducks were assumed to have been captured by Peregrines that were seen feeding on the carcasses. Of 39 ducks of known sex, 56% were males (Table 4.2). Ducks were hunted throughout the day. Individual falcons that could be recognized or were under continuous observation killed ducks in the morning and began attacking ducks again in late afternoon or evening. One adult falcon killed and partly consumed a duck around 06:00 and captured a second duck at 11:00.

Utilization of prey larger than teal was incomplete with only the fatty skin of the neck and 10–30% of the breast meat consumed after one meal. A Gadwall duck (*Anas strepera*), seized and killed in 10–12 cm of water, was abandoned after only part of the neck and back had been eaten. The falcon was probably unable to turn the submerged carcass over to obtain the breast meat. Migrating Peregrines killed ducks, fed for 15–35 minutes, and left again, soaring up high and disappearing from view in the distance. The remains of their prey were used by scavengers such as crows, hawks, and harriers. Peregrines that had remained in the area and kept under observation returned to their morning kill – if it was still available – for a second or third meal. Peregrines that killed ducks in late evening fed upon the carcass again early the next morning (Dekker 1984).

Food requirements of captive Peregrines are 120–150 gram/day of meat (Ratcliffe 1980). Brown and Amadon (1968) suggested one substantial kill might last large falcons in the wild for several days. Baker (1967) noted that wild, first-year Peregrines wintering in Britain killed two prey daily with a total weight of

>400 gram. My observations indicate that duck-hunting Peregrines wintering in British Columbia and migrating in Alberta killed one to three prey/day that weighed 260–700 gram each.

In British Columbia, four Peregrines were robbed of prey by Bald Eagles and two by Golden Eagles. All coastal Peregrines feeding on just-killed ducks were approached by Red-tailed Hawks (*Buteo jamaicensus*) or American Rough-legged Hawks (*Buteo lagopus*). Their attempts at taking over and holding onto the prey were usually warded off by large female Peregrines. Buteo hawks, Northern Harriers (*Circus cyaneus*), Glaucous-winged Gulls (*Larus glaucensus*), and on one occasion an immature Peregrine, were seen waiting near feeding adult female Peregrines. After the falcons had left, the scavengers moved in, but nearly all duck remains were eventually carried away by Bald Eagles.

Eagles are rare in central Alberta during the spring migration of Peregrines, but buteos are common and often rob male Peregrines. Sexual size dimorphism is pronounced in Peregrines and males are generally one-third smaller than females (Brown and Amadon 1968). After losing a just-caught Blue-winged Teal (*Anas discors*) to a Red-tailed Hawk, an immature male Peregrine captured another teal minutes later, which was taken over by a Swainson's Hawk (*Buteo swainsoni*). Both in British Columbia and in Alberta, I saw adult male Peregrines surrender just-caught pintail drakes to Red-tailed and Rough-legged Hawks. The Peregrines perched nearby and returned to the leftovers after the hawks had departed.

Female Peregrines routinely take prey from smaller Peregrines. Intraspecific klepto-parasitism by female falcons at the expense of the smaller sex is perhaps the reason why male Peregrines are much scarcer than females at coastal observation points where migrating Peregrines congregate (Dekker 1979). In Alberta, male Peregrines that captured small shorebirds or passerines transported them over long distances or consumed them while soaring after escaping from pursuing harriers, buteos, or other Peregrines (Dekker 1980). In British Columbia, I observed adult male Peregrines carry their prey to tall trees, which would give them a height advantage at the approach of potential klepto-parasites.

Peregrines did not transport ducks larger than teal unless the carcass was partly consumed. Peregrines that capture prey too heavy for carrying should be especially vulnerable to harassment by larger raptors, and this may have contributed to the employment of hunting strategies that reduce that risk. In British Columbia, Peregrines often seized ducks on or near the ground among marshy vegetation that obscured them from view. In Alberta, Peregrines that had been inactive during the afternoon began attacking ducks 0.5–1 hour after sundown. Crepuscular foraging activity by Peregrines has been reported by Beebe (1960), Clunie (1976), and Dekker (1980). Hunting at dusk or at nightfall can be expected to reduce the Peregrine's risk of klepto-parasitic conflict with eagles and other large diurnal raptors.

Table 4.1. Duck hunting techniques of Peregrine Falcons migrating in Alberta and wintering in British Columbia. Success rates are in brackets.

	Migrating in Alberta		Wintering in British Columbia	
	Hunts	Kills	Hunts	Kills
Ducks flying when attacked and seized (1 was struck down)	142	16 (11%)	19	3 (16%)
Ducks on the ground or in shallow water when attacked and seized	133	9 (7%)	24	7 (29%)
Totals	275	25 (9%)	43	10 (23%)

Table 4.2. Species and sex of ducks captured by Peregrine Falcons or on which they were seen feeding in Alberta (during migration) and British Columbia (during winter), 1969–1986.

Species	Male	Female	Unknown
Mallard	–	2	–
Gadwall	1	1	–
Northern Pintail	4	3	–
Green-winged Teal	5	2	–
Blue-winged Teal	2	–	2
American Wigeon	1	5	1
Northern Shoveler	3	3	–
Lesser Scaup	6	1	–
Unidentified	–	–	3
Totals	22	17	6

*Migrating Peregrines that kill ducks along the way consume only the skin of the neck and part of the breast muscle before resuming travel. However, if they make a late evening capture, they take a second meal the following morning. The leftovers serve a host of scavengers. On the west coast of Canada, wintering falcons are commonly robbed of their prey by eagles, which can force a Peregrine to kill more than one duck a day. This adult female, photographed by Jan Uilhoorn on the coast of California, was feeding on a Northern Shoveler (*Anas clypeata*).*

CHAPTER 5

Foraging habits of wintering Peregrines

Dekker, D. 1995. Prey capture by Peregrine Falcons wintering on Vancouver Island, British Columbia. Journal of Raptor Research 29:26–29.

Abstract – Peregrine Falcons (*Falco peregrinus*) wintering on southern Vancouver Island, British Columbia, killed 52 birds, of which 46 were ducks of five species. Twenty-four ducks were attacked and seized on land or in shallow water; 22 ducks were pursued in flight and borne down. Female Peregrines killed up to three ducks per day, but they were often robbed by eagles. Large Peregrines took prey from smaller conspecifics. Male Peregrines hunted mostly passerines. Male Peregrines that captured ducks larger than teal quickly lost them to klepto-parasitic buteo hawks. The most common hunting method was a low surprise attack initiated from a high tree perch, but the falcons also made long-range flights at ducks high in the sky.

INTRODUCTION, STUDY AREA, AND METHODS

The objective of this study was to closely observe and describe the hunting habits of Peregrine Falcons (*Falco peregrinus*) wintering on Vancouver Island, British Columbia. The study area was approximately 3 km^2 of agricultural fields surrounded by wooded hills near Victoria. Ponds and flooded fields, particularly after heavy rain, attracted several hundred ducks, mainly Mallards (*Anas platyrhynchos*) and American Wigeon (*Anas americana*). I traversed the area on foot and by vehicle. Peregrines were located by frequently scanning through binoculars or telescope and by noting alarm reactions of prey species. I minimized disturbance by watching the falcons from a distance, usually >200 m. Flying Peregrines were kept in view through binoculars to observe potential interactions with prey species. Falcons that were feeding on a kill were observed until they flew off. Prey remains were identified and examined but left in the field. I visited the study area on a total of 98 days in January and February of 1980–1994. Hours per day varied from one to nine. On most days I was in the area from sunrise to sundown, and I saw one or more Peregrines on all but three days. The earliest sightings occurred within half an hour after first light. A few falcons were seen at dusk, but most flew away near sundown, probably to distant roosting sites.

RESULTS AND DUSCUSSION

The study area was dominated by one evidently territorial adult female, which perched for hours on a large tree standing in the fields. She chased and repelled other females, adult or immature, but tolerated males. Each winter, I identified four or five different Peregrines of both sexes and belonging to different subspecies. Large adult females observed at close range looked like *F. p. pealei*, their white chests marked with dark spots (Beebe 1960; Palmer 1988). Other females with unmarked, salmon-coloured breasts seemed more typical of *F. p. anatum*. Some males looked small enough to belong to *F. p. tundrius* (White 1968). South of the study area, in Washington State, Anderson and DeBruyn (1979) captured and banded several wintering adult females, weighing 1181–1282 grams, which they identified as *F. p. pealei* and *F. p. anatum*, but they did not report any representatives of *F. p. tundrius*, and they saw very few adult males of any subspecies.

The total number of prey I saw captured by Peregrines in the study area was 52, of which 46 were ducks of five species: 19 American Wigeon, 13 Northern Pintails (*Anas acuta*), six Green-winged Teal, one Ring-necked Duck (*Aythya collaris*), one Bufflehead (*Bucephala albeola*), and six ducks of unknown species. Seventeen of these preys were male, 14 female, and 15 of unknown sex. Mallards were seldom attacked and never seen to be taken. Other prey species captured included two American Robins (*Turdus migratorius*), two Rock Doves (*Columba livia*), and one unidentified sandpiper.

Stealth hunts. Starting from a high tree perch, the falcons directed the majority of hunts at ducks sitting on wet ground or on the edge of water about 0.5 km or more away. The typical attack method was a fast flapping flight on a descending course. During the last stage of the attack the falcon sailed with flexed wings low over the ground or water. This stealth tactic was especially effective when the falcon's approach was screened by reeds or bushes. Ducks that detected the danger in time, flushed and raced for the nearest water where they were not attacked further.

After an unsuccessful hunt, the falcon returned to the same or another perch and resumed sit-and-wait hunting. Ducks that had been attacked several times became very wary and might be alerted well ahead of the falcon's approach by the rising of other ducks and gulls from adjacent fields. I hypothesize that the larger the number of birds present in the area, the greater the chance that the falcon was spotted soon after it left its perch, and the less chance it had of taking the target by surprise. If the intended prey flushed well ahead, the falcon did not complete its attack and returned to a perch. On some days, I saw a falcon make 10–15 surprise attacks over a period of 3–7 hours without making a kill.

Because of the frequency of aborted hunts, in which the falcon's approach was frustrated by early alarms, I have refrained from calculating a success rate. In an earlier paper that included some observations from this same study area (Dekker 1987), I reported a success rate of 21% ($n = 43$), which is close to the 26% ($n =$

57) rate recorded by Anderson and DeBruyn (1979). These success rates of wintering Peregrines are significantly ($P < 0.001$) higher than the 8.0% ($n = 1,125$) for migrating falcons in central Alberta (Dekker 1988).

Twenty-four ducks were seized very close to or on the ground, but in 11 of these captures the exact moment of contact was obscured behind rising ground, buildings, or vegetation. In all prey captures observed in detail, the Peregrine "bound" to its prey and held on. Not a single "knock-down" was recorded during this study, although the routine escape tactics of ducks – plunging into water and diving to dodge the attack – made it look as if an aerial hit had taken place. Knock-downs were also rare in a study of migrating Peregrines which included only one crippling aerial strike in 30 captures of ducks (Dekker 1987, and unpublished data).

Probably because of generally cool and wet conditions, Peregrines rarely soared in the study area, a common habit elsewhere (Dekker 1980, 1988). However, I saw two soaring falcons stoop at prey flying far below, one of which resulted in the capture of a Ring-necked Duck. Such hunting methods may be more common locally than reported here. Taking place at great altitudes, they easily escape notice.

Aerial pursuits. During periods of low water there were few ducks on the fields and the falcons concentrated on intercepting prey that happened to fly by. Overtaken from behind, ducks were seized in the air and borne down. Meeting some ducks head-on, the falcons approached low over the ground and grabbed their prey in a steep upward swoop. An adult pealei male made long climbing flights at groups of ducks approaching high over the hilly skyline and still >1 km away. As the falcon closed in, the flocks scattered and veered aside. Selecting its target, the falcon pursued the ducks while they descended in an attempt to reach water or other cover. A teal was seized low over the ground. Three pintails were killed after they had landed, respectively, on a wet field, in tall grass, and on a frozen pond.

Attacks on flying ducks resulted in 19 aerial captures. Three pintails or wigeon were seized above water, but released again after the falcon had lost altitude and was apparently unable to reach dry land. One duck was hanging upside-down in the falcon's clutches before it was dropped and splashed into the water. Too heavy to carry, ducks the size of wigeon were plucked and eaten where they fell, except in one incident when a female falcon brought a wigeon down on a frozen pond and dragged it 80 m over the smooth ice to the shore. Falcons had no difficulty carrying teal, which were transported over long distances and consumed on high tree perches

All falcons, including males, seemed eager to pursue teal, probably because it is easily killed and carried out of reach of scavengers. One large female hunted and killed only teal, ignoring all other species of duck during 15 consecutive days of observation (See Dekker 1999).

Peregrines of both sexes sometimes hunted together. On two occasions a male and female pursued the same Green-winged Teal (*Anas crecca*). One of these was caught by the male, who was then immediately set upon by the female and followed out of sight. In another chase, an adult male pursuing a teal was joined by a smaller immature male. The adult caught and consumed the prey. Less than an hour earlier, this same adult male had robbed the immature male of a just-caught American Robin. However, the adult did not eat the robin and left the carcass on a pole when he took off to hunt the teal. (Evidently, this Peregrine preferred duck!)

Although immature male Peregrines often pursued teal (with one capture seen), they hunted mostly passerines, particularly American Robins and Starlings (*Sturnus vulgaris*). Of two flying passerines that were met head-on, one was seized directly; the other tried to dodge the attack by plunging steeply down, to be overtaken and seized by the stooping falcon. One robin was taken after it flushed from the ground just ahead of an immature male peregrine that had approached the robin in a long-range surprise attack initiated from a tree perch. Male Peregrines made three prolonged pursuits of sandpipers, of which one was successful.

Domestic pigeons and feral Rock Doves were common in the area and sometimes attacked, but rarely captured. One courting dove, sailing over a farmyard, was seized directly from behind by an adult female falcon in a fast surprise attack taking place between trees and buildings. Another dove was caught after a twisting pursuit by an adult male Peregrine. Upon examination of the remains, it turned out that the dove's crop was full of grain, which must have hindered its mobility and increased its vulnerability.

Interrelationship with other raptors. Male Peregrines migrating in central Alberta consumed sandpipers while soaring at high altitudes so as to escape from klepto-parasitic larger raptors (Dekker 1979, 1980). Eating-on-the-wing was not seen during this study, but male Peregrines always took small prey to a high tree perch, probably as a precaution, giving them a height advantage if pirates should turn up. Two males that had captured ducks too large to carry were robbed by Red-tailed Hawks (*Buteo jamaicensis*). In one of these incidents, the red-tail was eventually flushed by a Rough-legged Hawk (*Buteo lagopus*), while the Peregrine waited in a nearby tree. He returned to his kill (a pintail drake) after both buteos had departed.

Female falcons feeding on ducks ignored Glaucous-winged Gulls (*Larus glaucescens*) or Northern Harriers (*Circus cyaneus*), but they reacted defensively at the approach of Red-tailed or Rough-legged Hawks. If needed, the hawks were driven off with loud calls and aggressive swoops. However, the falcons were powerless against Bald Eagles (*Haliaeetus leucocephalus*) and Golden Eagles (*Aquila chrysaetos*).

Piracy by eagles had a serious impact on Peregrine foraging behaviour, forcing the falcons to kill ducks in excess of their own food requirements. For instance, on 21 January 1993, from 08:00 to 17:15, I watched an adult female Peregrine make 20 attacks (launched from the same tree perch) on ducks feeding or resting on wet meadows. That day she captured three wigeon, but two of these were almost immediately lost to Bald Eagles, respectively at 09:00 and 16:15. The third kill was made at 16:50, just before dusk. This time no eagles showed up and the falcon fed until 17:20. In total, I saw Bald Eagles pirate ten ducks shortly after they had been killed by Peregrines. In addition, three ducks were commandeered by Golden Eagles. One of these Golden Eagles subsequently lost the carcass to an adult Bald Eagle. A similar dynamic between klepto-parasites and Peregrines was reported from Washington State, where a wintering female falcon killed three ducks in one day, but lost the first two to pirates and scavengers (Anderson and DeBruyn 1979).

Gyrfalcons (*Falco rusticolus*) elicited fierce defensive behaviour from Peregrines. An immature male Gyrfalcon took a just-caught pintail duck from an adult female Peregrine, which perched in a tree and returned to the kill after the larger falcon had fed and left.

Peregrines sometimes took advantage of disturbances among waterfowl caused by other raptors or people. At the approach of harriers or eagles, ducks flew up from floodwater, which increased their vulnerability to falcons. On three occasions, I saw Peregrines capture teal that had flushed to evade Bald Eagles. In a similar way, Peregrines captured two pintails, two teal, and one Bufflehead involuntarily flushed from ponds or ditches by me or other birdwatchers.

Kill frequency. Anderson and DeBruyn (1979) assumed that Peregrines killed and ate twice daily, since radio-tagged falcons observed with a full crop in the early morning hunted again in the evening. This assumption may be false, since two daily periods of hunting do not automatically mean that two kills were made. It is possible that falcons hunt unsuccessfully during the evening and enter the night hungry, to resume foraging at daybreak. Conversely, falcons that do make a late night kill might eat from the leftovers the following morning. In Alberta, I saw falcons return at first light to their previous night's duck kill (Dekker 1984). During this study, a Peregrine picked up the remains of a duck it had caught and partly consumed the day before, but scavenging gulls had already removed all edible tissue. Another adult female landed and fed on the flattened carcass of an American Coot (*Fulica americana*) lying on the pavement of a county road. I do not know whether or not the coot had actually been captured and killed by the falcon before it was driven over by a vehicle. Most Peregrine prey remains were utilized quickly by scavengers and eventually carried away to distant trees by eagles. The result is that Peregrines rarely get a chance to eat twice from the same kill.

Duck kills occurred throughout the day but most often between 10:00 and 12:00, with nine kills each hour. During the early morning, prior to 10:00, I recorded four to five captures each hour. Hunting activity was lowest during the afternoon averaging only three captures per hour. The frequency climbed to five or six kills for the last two hours of the winter day.

Figure 5.1. In a typical trade-off between safety and food, wintering wigeon that spend the day on flooded meadows have to leave the security of the water to feed on green grass. After the closest pond margins become denuded, the ducks have to venture out ever farther, thereby increasing the danger of being surprised by a low attack of a Peregrine. Apparently, the hungrier and more careless the duck, the greater its risk.

CHAPTER 6

Marine Peregrines hunting seabirds

Dekker, D. and L. Bogaert. 1997. Over-ocean hunting by Peregrine Falcons in British Columbia. Journal of Raptor Research 31:381-383.

Abstract – Peregrine Falcons (*Falco peregrinus pealei*) on breeding territory in the Queen Charlotte Islands, British Columbia, made 73 hunts over the ocean of which 16 were successful (22%). The most common hunting method (80%) was a stealth attack low over the water to take swimming seabirds by surprise. Most hunts were launched from high tree perches. Falcons also made long-distance search flights, including soaring. The majority (88%) of prey captured was small alcids, such as murrelets and auklets. Ten were struck on the surface of the ocean and retrieved after one or more return passes. In addition, two flying alcids was grabbed in the air, and two phalaropes were seized on the water or just after they flushed. On eight occasions, Peregrines with prey were pursued by Bald Eagles (*Haliaeetus leucocephalus*), which pirated two preys.

INTRODUCTION

The breeding population of Peale's Peregrine Falcons (*Falco peregrinus pealei*), nesting on Langara Island on the north Pacific coast of British Columbia, Canada, has been monitored for more than half a century (Beebe 1960; Nelson 1990). The collection of prey remains near nests indicates that these falcons prey nearly exclusively on small seabirds, particularly murrelets, auklets, and petrels. Beebe (1960) speculated that the falcons hunted Ancient Murrelets (*Synthlboramphus antiquus*) mainly at dusk and dawn when the murrelets fly to and from nesting colonies in island forests. Peale's Peregrine Falcons breeding on the Aleutian Islands off Alaska also prey on small seabirds, but little is known about their foraging methods (White 1975; Sherrod 1988). To fill this gap in knowledge about the hunting strategies and success rates of these marine Peregrines, we visited Langara in 1995 and 1996. This paper reports on 73 observed hunts.

STUDY AREA AND METHODS

Langara Island is at the northwest tip of the Queen Charlotte Islands off the coast of British Columbia, Canada. The island is about 10 x 6 km and mainly forested with low hills. Coastal cliffs are used by Peregrine Falcons as nesting sites (Beebe 1960). Our observations were made after the nesting season was over, from 30 August to 6 September 1995 and from 7 to 22 August 1996. During 22 days we spent a total of about 200 hours atop a 30 m high shoreline cliff. Primarily, we observed one territorial pair that nested on a forested point jutting out into the ocean 1.3 km to the west of our observation point. The falcons often perched on the branches of dead conifers that protruded above the shoreline forest, or on a 4–6 m high rocky islet just offshore. Another breeding pair occupied a set of cliffs just east of our vantage point, but their perches were blocked from our view by trees. We spotted this second pair of falcons only by chance if they flew by our post. We also saw several first-year immature falcons that had recently become independent.

We watched flying Peregrines through binoculars or a 20x wide-angle telescope for as long as they remained visible. Perched falcons were monitored continuously, often for periods of up to four hours, in the hope of seeing them take off on a hunt. To the west and north, we had an unobstructed view across a wide bay and open ocean. Usually, hundreds of alcids dotted the water and small flocks or single birds flew very low (≤ 1 m) over the waves. They included several species, such as Cassin's Auklet (*Ptychoramphus aleuticus*), Ancient Murrelet, and Marbled Murrelet (*Brachyramphus marmoratus*). With mean weights of 170 to 225 gram, these alcids are common prey of Langara Peregrines (Beebe 1960; Nelson 1970).

A hunt consisted of one completed attack a Peregrine made on a target prey species, including one or more capture attempts. We use the terms hunt and attack interchangeably (Dekker 1980). A foraging flight could contain one or more hunts (Buchanan 1996).

RESULTS

We observed a total of 73 hunts of which at least 16 were successful (22%). During 13 of these hunts we briefly lost sight of the falcon over the ocean; 11 returned without prey to their perch, but two falcons were pursued by a Bald Eagle (*Haliaeetus leucocephalus*), so we assumed that they were carrying prey. Sixty hunts were observed in detail from start to finish.

Still-hunting. Fifty-eight hunts (80%) were launched from high tree perches and the intended target was usually <1.5 km from shore. Flying out at 50–100 m over the ocean, the falcons gradually descended low over the water, beating their wings for some distance until they glided with wings partly flexed. At the conclusion of the attack, the falcon abruptly pulled up and headed back to land, with or without a prey. Particularly during windy conditions the falcons perched on the islet and took off low, darting between the waves. Some capture attempts were made just beyond a large, partly submerged rock that appeared to conceal their approach

Long-range search flight. During longer and higher flights, the falcon might suddenly change direction and descend at increasing speed very low over the water. We saw 15 (20%) such hunts. After an unsuccessful capture attempt, these falcons did not immediately return to land, but disappeared from view in the distance, probably continuing their foraging flight. On six occasions, the falcons soared, either over shoreline hills or water. At 07:00 on a warm day, an adult male circled over the ocean for about 10 minutes, then flew down obliquely to pursue a low flying alcid that dived into the water. On another early morning, an adult female tried to soar but gave up soon and resumed flapping flight, vanishing far away over the skyline. Long-distance search flight over the ocean, including soaring during suitable weather, seemed more typical of immatures that were often chased away from the shoreline by territorial adults.

Attacks on swimming alcids. At least 39 hunts, of which ten (26%) were successful, were directed at swimming alcids. Through the scope we could clearly see that most targets dived under water at the last possible moment. The attacking falcon then passed low over the spot, pulled up and returned to shore. Alcids spread their wings to dive, which results in a visible splash in the water. Sometimes, we saw two or three splashes, indicating that more than one bird dived to avoid the falcon. In one hunt, an alcid near the crest of a rising wave submerged in time to avoid capture, while another, in the trough of the next wave and also directly in line with the falcon's low approach, failed to dive and was struck.

On impact, the falcon dipped down, evidently raking the alcid with its claws, then shot upward and quickly doubled back to retrieve the disabled prey from the water. In six out of nine cases, the falcons seized the alcid after a single pass; in three attacks they made two or three repeat passes. Multiple passes occurred during calm conditions, when the falcons apparently had more difficulty than on windy days braking their speed enough to retrieve the wounded or dead prey. In one closely watched hunt the falcon failed in three tries while the crippled prey dodged, thrashing about on the water. Finally, it was picked up by the first of three approaching adult Bald Eagles. In another unsuccessful capture attempt, a falcon briefly hovered over the water, possibly waiting for the reappearance of an injured alcid that had dived below the surface. It probably had been hit but not hard enough to prevent it from diving. In one surprise attack, an adult male, gliding fast and low over the calm water, seized a small alcid directly from the surface and carried it away.

Attacks on low-flying alcids. At least 12 hunts were directed at alcids flying <1m over the water, with two (17%) captures. Ten of these flights originated from perches and two from high cruising flight. To evade approaching falcons, flying alcids dived at once into the water. During one successful attack by an adult female falcon, the alcid was seized directly out of a small flock while the others splashed down. Another flying alcid was captured by an immature female falcon that soared for many minutes over the ocean, until it stooped and raced low over the water for a distance of several hundred metres to overtake and seize the prey.

Other prey species. We observed nine attacks on swimming phalaropes (*Phalaropus* spp.), a species that breeds in arctic regions and was passing through the area on migration. Two of these hunts were successful. Phalaropes that flushed in time were briefly pursued. One was seized either on or just over the water by a falcon darting low between choppy waves. The other phalarope was taken by an adult male in a typical surprise attack. Gliding along the surface, he grabbed his prey directly from the water out of a small flock that flushed just too late.

Klepto-parasitic eagles. Eight Peregrines with prey were closely pursued by one to three adult Bald Eagles. Several falcons carrying a just-caught prey climbed high over the ocean, describing a wide half circle, before returning to land. It was our impression that these falcons wanted to reach sufficient speed and altitude so as to get a head start on possible pursuers. One adult female falcon appeared unable to gain height quickly enough. After a persistent and very close pursuit by an adult eagle, she happened to closely pass by our observation point, and perhaps startled by us, released her prey over our heads. By contrast, an immature falcon, carrying its prey over the open ocean, successfully dodged about 20 close passes by an adult eagle until it eventually gave up the pursuit. Some Peregrines took their catch into narrow cliff ravines where eagles might have difficulty following. Other falcons consumed their prey on tall trees, giving them a height advantage at the approach of pirates.

Prey captures were made throughout the day, but most often (37%, $n = 27$) between 09:00–12:00. Six falcons were observed with prey between 06:00 and 09:00, five during 12:00–15:00, and three each during 15:00–18:00 and 18:00–21:00.

DISCUSSION

Our study indicates that maritime Peregrines make "contour-hugging flights low over the ocean, using waves to conceal their approach until they come suddenly upon surface-swimming water birds which are panicked into flying or diving" (Cade 1982:62). However, we saw nothing to support claims that these Peregrines learn to wait out the dives and return to attack again when the birds have surfaced and are exhausted for oxygen. Our observations also differ from reports to the effect that *F. p. pealei* habitually strikes flying alcids an aerial blow that causes them to fall into the water (Sherrod 1988). On the contrary, the Peregrines in our study captured low-flying alcids by seizing them directly in their feet and carrying them to land. This is also the common method employed by Peregrines hunting ducks and shorebirds in Alberta and British Columbia (Dekker 1980, 1995). Likewise, British Peregrines hunting over the ocean captured the majority of prey by "binding" to them in mid flight (Parker 1970; Treleaven 1980)

In this study, the dominant hunting method was a low stealth attack with the objective of taking the prey by surprise. This is also the major strategy of migrating Peregrines in Alberta (Dekker 1980, 1988). Peregrines wintering on the coast of Scotland used surprise in 36% of hunts (Cresswell 1996).

Perch-and-wait hunting, similar to the method reported in this study, is the most common and successful hunting method of Peregrines worldwide (Palmer

1988). It is used routinely by falcons that nest on cliffs overlooking large expanses of water (Treleaven 1980; Bird and Aubry 1982), or that perch on high trees lining wetlands (Dekker 1995). By contrast, in central Alberta where high perches are scarce, migrating Peregrines mainly employ active search methods including flapping flight and soaring (Dekker 1980, 1988). In this study, long-range hunting flights over the ocean appeared to be typical of immature falcons, perhaps because they are chased away from shore by territorial adults. In areas away from occupied nesting cliffs, perch-and-wait hunting might well be the preferred method of immatures as well as adults.

Our observed hunting success rate of 22% falls within the range of values from other regions of the world (Roalkvam (1985). The only data for Peregrines hunting over the ocean are from Britain, where Peregrines specialize on Rock Doves (*Colomba livia*) with success rates ranging from 16% (Parker 1979) to 52% (Treleaven 1980). To our knowledge, our study is the first to include success rates for Peregrine Falcons hunting seabirds, and the first to report that these marine Peregrines commonly strike swimming prey on the water. This capture technique was only once observed by Nelson (1970). There is also one old anecdote quoted in Bent (1937:55) of a low flying Peregrine striking a small grebe dead on the surface of the ocean off the coast of Vancouver Island.

The frequency of klepto-parasitic attacks by Bald Eagles observed in this study may explain why Langara Peregrines do not capture the locally common Pigeon Guillemot (*Cepphus columba*), which weighs about 450 gram and is twice as heavy as the smaller alcids. During 26 years of research on Langara Island, R. W. Nelson (personal communication) did not find any guillemot prey remains in Peregrine nests. In contrast, legs and wings of the very similar Black Guillemot (*Cepphus grylle*) are quite common in Peregrine nests along the Arctic Ocean (Bradley and Oliphant 1991) where Bald Eagles are absent. Although the arctic race of the Peregrine (*F. p. tundrius*) is smaller than *F. p. pealei* (White 1975), it is apparently quite capable of carrying guillemots. However, at Langara, the weight of a guillemot can be expected to even more increase the vulnerability of Peregrines to piracy by Bald Eagles. The risk of losing prey to klepto-parasites also affected choice of prey and hunting habits of migrating Peregrines in Alberta (Dekker 1980). As well, on Vancouver Island wintering Peregrines, males as well as females, showed a strong preference for light-weight prey such as Green-winged Teal (*Anas crecca*), which can be carried out of reach of pursuing eagles (Dekker 1995).

CHAPTER 7

Over-ocean flocking by wintering Dunlins

Dekker, D. 1998. Over-ocean flocking by Dunlins (*Calidris alpina*) and the effect of raptor predation at Boundary Bay, British Columbia. Canadian Field-Naturalist 112:694–697.

Abstract – Flocks of Dunlins (*Calidris alpina*) often exceeding 10,000 birds, wintering at Boundary Bay, British Columbia, remained airborne over the ocean for the duration of the high tide period when their intertidal feeding grounds and open roosting habitats were inundated. I postulate that these flights, often lasting 2–4 hours, are an anti-predator strategy to avoid surprise attacks by raptors over the saltmarsh shore zone. Peregrine Falcons (*Falco peregrinus*) attacked Dunlins 302 times with 28 captures, a success rate of 9.3%. Surprise hunts over the saltmarsh had a success rate of 33.0% in 15 hunts, significantly higher ($P < 0.05$) than the 8.0% in 287 hunts over the open mudflats. All 29 hunts by Merlins (*Falco columbarius*) began as surprise attacks although all five kills were made after long pursuits.

INTRODUCTION

Shorebirds wintering or staging on coastal estuaries mass together during high tide on traditional roosting sites that remain above water (Morrison 1977; Hicklin 1987; Mawhinney et al. 1993). In Washington State, wintering Dunlins (*Calidris alpina*) fly to wave-swept beaches that can be as far as 15 km from the estuaries (Buchanan 1996). In Holland, Knots (*Calidris canutus*) roost on open mud banks 7.5 km away from their feeding grounds (Piersma 1994). In California, wintering Dunlins and other sandpipers collect on shoreline pastures when the adjacent lagoon becomes inundated (Page and Whitacre 1975). A portion of the Dunlins wintering in coastal British Columbia are known to roost on inland fields (Butler 1994).

Open-country birds flock together as a response to predation. Individuals in large flocks have a lower chance of being killed by an attacking raptor, and they spend significantly less time in surveillance and more time feeding than do individuals in small flocks or lone birds (Powell 1974; Lima 1987). Dense flocks might also discourage predators from entering the flock for fear of damaging themselves (Tinbergen 1951). At the approach of predators, flocks of shorebirds engage in tightly synchronized flight manoeuvres (Buchanan 1996). Falcons selectively hunt lone and juvenile shorebirds (Dekker 1980; Kus et al. 1984; Bijlsma 1990; Warnock 1994). This paper reports on the roosting behaviour of Dunlins and their interaction with predators at Boundary Bay in southwestern British Columbia.

STUDY AREA AND METHODS

Boundary Bay is part of the Fraser River delta (49°05'N, 123°00'W) in southwestern British Columbia. The bay is 16 km across and the intertidal zone is 4–5 km wide at the lowest ebb. During January, daytime high tides reach over 5 m above lowest normal tides (Point Atkinson chart) and inundate all mudflat habitats including most of the narrow strip of saltmarsh. The bay is contained by a low dike and the adjacent delta is mainly farmland. Rich in marine foods, Boundary Bay is a major migration stopover for shorebirds and a wintering ground for 25,000–35,000 Dunlins (Butler 1994; Butler and Kaiser 1995). Raptors are common. For a more detailed description of the Fraser River delta and its avian inventory see Butler and Campbell (1987).

During January 1994–1998, I spent 54 days, from first light to dusk, in the study area. Especially during windy and rainy weather, I watched from a parked car. At other times I walked on the dike. I roughly estimated Dunlin numbers by counting them on sections of shore and extrapolating the totals during outgoing tides when the birds spread out over the mudflats. I frequently scanned the area through binoculars to discover alarm behaviour of Dunlins and to spot predators. Flying falcons were watched until they flew out of sight. Perched falcons were monitored through binoculars or a telescope in the hope of seeing them leave on a foraging flight. The term "hunt" means one completed attempt at capturing a Dunlin of which the outcome was known. This conforms to the definition used by other researchers who presented large data sets on raptors hunting shorebirds (Page and Whitacre 1975; Dekker 1980, 1988; Cresswell 1996).

RESULTS

High-tide behaviour of Dunlins. Each January I estimated total Dunlin numbers in Boundary Bay at circa 30,000. When the flood tide inundated the mudflats, the Dunlins assembled along the edge of the saltmarsh. As the tide crested, many birds stood up to their bellies in water before they finally flew away. Some flocks departed inland, particularly after heavy rain when fields and meadows became waterlogged. However, the majority of the Dunlins flew out over the ocean, where they coursed back and forth. The flock drifted in a loose cloud on the wind or coalesced into a dense, undulating stream low over the waves. The Dunlin flocks frequently split up and rejoined again later or merged with other flocks.

I first noted this flight behaviour in the late afternoon of 23 January 1994, my first full day on the dike, when I watched the flocks for 1.5 h and did not see them return to land. During subsequent observations, these over-ocean flights proved to be routine. Depending on the timing of the tides, I often saw the entire high-tide flocking intermezzo including the birds' return 2–4 hours later. The time span appeared to vary with the height of the tide and the extent to which the saltmarsh had been inundated.

While the flocks were airborne over the ocean, their flight pattern was influenced by weather conditions. With light winds, the birds cruised about slowly with fluttering wings in loose aggregations that looked from a distance like veils of fog or smoke. During rain or when strong winds were blowing, the flocks were generally much smaller and flew very low over the waves, making slow progress and often dipping out of sight. On some stormy days no birds could be seen with the unaided eye. It is not impossible that they had departed for some distant roosting site. However, scanning the ocean through binoculars, I usually picked up one or more flocks estimated to be >2 km offshore.

When the tide began to ebb, flocks of Dunlins returned over the water or flew upwind along the shore. Landing tentatively on the first mudflats to emerge, the birds soon spread out into the shallows to feed. In the meantime, smaller flocks returned from inland roosting sites.

Predator interaction with Dunlins. During this study, I identified 11 species of diurnal birds of prey and two owls, but only five of these commonly hunted Dunlins. The most numerous raptor was the Northern Harrier (*Circus cyaneus*), and I recorded many sightings each day. Harriers frequently approached feeding or roosting Dunlin flocks in a low level sprint intended to take a prey by surprise. Once the Dunlins were airborne, harriers did not pursue them for more than a few seconds, and the Dunlins descended again very soon, often returning to the same place. Yet, persistence paid off; I saw three captures of Dunlins by harriers.

Each winter there were at least five or six different Peregrine Falcons (*Falco peregrinus*), recognizable by differences in size and plumage, foraging over the bay. I saw one or more each day and recorded a total of 302 hunts aimed at Dunlins including 28 captures. Hunting Peregrines caused violent panic responses among the Dunlins, which mounted high into the sky and formed dense globular flocks. After one or more attacks, the flocks abandoned the immediate area and descended >1.5 km away.

Dunlins reacted in a similarly defensive manner to hunting Merlins (*Falco columbarius*). Although there probably was more than one Merlin in the area, I did not see them each day. I recorded 29 Merlin attacks on Dunlins, resulting in five kills. Merlins as well as Peregrines employed a wide variety of hunting methods, including low surprise approaches, multiple stoops at flying flocks, and long pursuits of single Dunlins.

A locally abundant potential predator of Dunlins was the Glaucous-winged Gull (*Larus glaucescens*). It had little chance of catching a Dunlin on open mudflats, but it was opportunistic. Once I saw a swimming gull, head held low, approach and suddenly lunge at Dunlins feeding along the edge of the saltmarsh.

After watching half a dozen tries, I just missed the moment of capture, but I saw this gull swallow a Dunlin, head first, and vigorously flapping its wings. This incident happened in January 1998 during a spell of extreme cold when the Dunlins were probably hungry and less alert to danger.

Bald Eagles (*Haliaeetus leucocephalus*) were numerous (100–200) and attempted to rob Peregrines and Merlins carrying prey. I saw one eagle pursue a Peregrine until it dropped its just-caught Dunlin. Eagles, as well as gulls and harriers, also joined chases in progress, intent on picking up Dunlins that were struck down by the falcon or after they took cover in water or vegetation. The eagles succeeded in three of these pursuits. The loss of prey forced the falcons to hunt again. Another persistent klepto-parasite was a Prairie Falcon (*Falco mexicanus*), which wintered in the area in 1997. It twice attempted to rob a Peregrine of a just-caught Dunlin, once succeeding.

Peregrines and Merlins hunted throughout the day, even at low tide when the mudflats were >2 km wide and the Dunlins nearly out of sight from the dike. Flocks that were cruising over the ocean at high tide were also attacked. However, I saw falcons most often just before or after high tide when the Dunlins were massing near shore, and appeared to be very nervous, frequently flushing at the approach of harriers. An attack by a Merlin or Peregrine, streaking at high speed over the dike and the vegetation of the saltmarsh, would result in the flock's departure and trigger the high-tide flocking intermezzo.

The number of times Peregrines were seen to make surprise attacks over the saltmarsh was relatively small ($n = 15$) because the Dunlins spent little time there. But five attacks ended in capture, a success rate of 33.3%, significantly higher ($G = 4.83$, $P < 0.05$) than the 8.0% in 287 hunts over the open mudflats and the ocean.

I observed several surprise attacks by Merlins, low over the saltmarsh, but the outcome remained uncertain because the Merlin soon was hidden from view among flushing Dunlins. I spotted two Merlins carrying prey away over the saltmarsh, after I had missed the initial phase of these hunts. All 29 Merlin hunts observed from their start to their conclusion began as an attempt to take a Dunlin by surprise, even over open mudflat habitat. All of these 29 hunts failed. Subsequently, these Merlins changed tactics and concentrated on flocks over the mudflats well out from shore. All five captures were the result of long pursuits of single Dunlins that had been isolated from the flock. For a more detailed account of Peregrine and Merlin foraging habits see Dekker (1999).

DISCUSSION

The 9.3% hunting success of Peregrines in this study is nearly equivalent to other studies, but the 17.2% success rate of Merlins, although the sample is small, is higher than rates reported elsewhere (Table 7.1). The most common hunting strategy used by Peregrines and Merlins was surprise, which was also the main method of falcons hunting shorebirds elsewhere (Page and Whitacre 1975; Dekker 1980, 1988; Cresswell 1996). Nevertheless, in this study most captures by Peregrines (20 of 28) resulted from attacks on flying flocks and long pursuit of single Dunlins. All five kills by Merlins were the result of long pursuits, after attempts at surprise had failed. As Dunlins avoided the shore, the falcons were forced to hunt over the open flats where surprise was more difficult to achieve. In this study, I observed many different individual Peregrines and their hunting styles varied. Some adults hunted exclusively by surprise, whereas immatures varied their methods and often pursued and repeatedly swooped at a single Dunlin until it was either caught or escaped into cover.

Dunlins, like many other small shorebirds, do not like to sit in vegetation because it blocks their view and predators are quick to take advantage of cover to maximize the element of surprise (Dekker 1980, 1988). Therefore, I postulate that over-ocean flocking, as observed in this study, is an anti-predator strategy. Dunlins at Boundary Bay had to resort to over-ocean flocking because they could not locate suitable, bare ground for roosting after all of the intertidal zone had been inundated by the high tides. I hypothesize that these Dunlins, instead of remaining airborne over the ocean, would have roosted on open sites if these had been available. They do so in adjacent Washington State (Buchanan 1996). Interestingly, at Tsawwassen, 5 km northwest of Boundary Bay, several thousand Dunlins mass together at high tide on the gravel tip of the breakwater jetties of the ferry terminal (Rick Swanston, personal communication). In addition, as mentioned, a portion of the Dunlins flies inland at high tide. Their number can be substantial. Butler (1994) counted a January maximum of 13,400 on fields in the Fraser delta. Roosting on fields and ferry jetties can be considered as a relatively recent adaptation, since such opportunities were absent prior to European settlement of the area, about a century ago, when the entire delta was likely covered with dense vegetation. It will be interesting to see whether inland roosting increases or decreases in future.

The metabolic cost to shorebirds of flights to and from distant roosting sites is considerable (Piersma 1994). The expenditure of energy during the over-ocean flocking phenomenon reported in this study is probably only possible if the birds have access to an abundance of food. This appears to be the case; Boundary Bay is exceptionally rich in marine organisms (Butler 1994).

It is interesting to observe Dunlin behaviour when and where these food resources are less accessible. For instance, I saw no over-ocean flocking during an extreme cold spell in January 1998, when most of the intertidal zone at Boundary Bay was covered with ice. Small flocks of Dunlins that remained in the area stood on the ice far from shore. Other small flocks spent the high-tide period searching for food in the saltmarsh. After the tide dropped, these Dunlins moved back onto the mudflat, feeding ravenously. Astonishingly, they showed minimal reaction to hunting Peregrines and Merlins. Instead of taking to the air in defensive formations, only those Dunlins directly in line with the approaching falcon got out of the way, to quickly land again. I hypothesize that these Dunlins had no energy to waste. The need to eat and the need to avoid predation is a trade-off. After the cold spell abated, Dunlins quickly returned to former numbers and over-ocean flocking resumed. I saw no over-ocean flocking during two days in April 1998 when the daytime high tides remained about 1 m below January levels. Then, the Dunlins and other sandpipers stayed on the mudflats away from shore all day.

The significance of the phenomenon reported in this study, that many Dunlins fly out over the ocean during the high-tide interval in January, when the entire intertidal zone is inundated by peak tides, has to my knowledge not been recognized before. However, high-tide flights of Dunlins have been observed at several locations along the Atlantic coast of Canada, namely at the Bay of Fundy, but their significance has neither been understood, nor has the phenomenon been singled out for publication (Tony Erskine, personal communication).

Table 7.1. Success rates of Peregrine Falcons and Merlins hunting small shorebirds.

Hunts	Captures	% Success	Source
Peregrines			
302	28	9.3	This study
233	20	8.6	Cresswell 1996
569	50	8.8	Dekker 1988
Merlins			
29	5	17.2	This study
182	21	11.5	Cresswell 1996
343	44*	12.8	Page and Whitacre 1975
223	28	12.6	Dekker 1988

*Includes a small but unspecified number of passerines.

Addendum: The duration of over-ocean flocking. In 2006 and 2008, for three periods of respectively 20, 12, and 5 full days, I monitored the high tide behaviour of Dunlins wintering at Boundary Bay. Over-ocean flocking (OOF) was observed on 28 of the 37 days. The details are as follows.

During the 20 day period from 9 to 28 January 2006, I recorded OOF on 18 dates. The duration ranged from 1.5 to 6.5 hours, with a mean on 3.2 hours/day. On two days the Dunlins did not fly at all and instead roosted on the outside edge of the saltmarsh. On one of these days, the weather was calm with heavy rain throughout the high tide period. The second day was clear and completely calm. This points to the importance of wind to facilitate OOF and minimize its energetic cost. As postulated in the paper above, Dunlins engaging in OOF make use of light to moderate winds to float, as it were, on the lateral flow of air. For a graph detailing the day-to-day flight times during this 20-day period, see Ydenberg et al. 2009.

I again observed the high tide behaviour of the Dunlins wintering at Boundary Bay from 4 to 15 February 2008. Weather conditions were generally rainy and windy, and the high tide point fell in the early morning before or soon after my arrival in the study area. OOF occurred on five of 12 dates, but it was intermittent and lasted less than two hours. On four dates, most of the Dunlins roosted on the foreshore during the high tide interval or flew inland to forage on flooded meadows. Several thousands of Dunlins remained there for part of the day even after the tide had begun to recede. This indicates the importance for Dunlins of temporary inland feeding grounds that may become attractive after rain.

On each of the five days from 20 to 24 October 2008, OOF varied in duration from 2.5 to 4.0 hours, with a mean of 3.3 hours/day. However, on all of these days large flocks (>500) stayed behind and roosted on the edge of the marsh, flying out intermittently over the ocean. After the tide began to recede, the Dunlins formed dense rafts in the shallows and roosted for another hour or more, indicating that these Dunlins were not yet hungry enough to start feeding.

*Adult male Peregrine with captured Dunlin (*Calidris alpina*) at Boundary Bay, Canada. (Photo: Miechel Tabak*

*The vast majority of Pacific Dunlin (*Calidris alpina*) wintering on the coast of British Columbia spends the high tide interval in flight over the ocean. Some flocks depart well before all intertidal mudflats are inundated; others crowd together in the shallows until they are flushed by an attacking Peregrine or Merlin. The duration of over-ocean flocking flights was timed at 1.5 to 6.5 hours. (Photos: Brian Genereux)*

CHAPTER 8

Peregrine predation on Dunlins in relation to ocean tides.

Dekker, D. and R. Ydenberg. 2004. Raptor predation on wintering Dunlins in relation to the tidal cycle. The Condor 106: 415–419.

Abstract – At Boundary Bay, British Columbia, Canada, Peregrines (*Falco peregrinus*) captured 94 Dunlins (*Calidris alpina*) in 652 hunts. The two main hunting methods were open attacks on flying Dunlins (62%) and stealth attacks on roosting or foraging Dunlins (35%). Peregrines hunted throughout the day, yet the kill rate per observation hour dropped 1–2 hours before high tide and peaked 1–2 hours after high tide. The drop in kill rate coincided with the departure of the mass of Dunlins for over-ocean flights lasting 2–4 hours. The peak in kill rate occurred just after the tide began to ebb and the Dunlins returned to forage in the shore zone. The hypothesis that closeness to shoreline vegetation is dangerous for Dunlins is supported by three converging lines of evidence: (1) the high success rate (44%) of Peregrine hunts over the shore zone compared to the rate (11%) over tide flats and ocean; (2) the high kill rate per observation hour at high tide; and (3) the positive correlation of kill rate with the height of the tides. Seven of 13 Dunlins killed by Merlins (*Falco columbarius*) and all five Dunlins killed by Northern Harriers (*Circus cyaneus*) were also captured in the shore zone.

INTRODUCTION

Many researchers have implicated predation risk as an important determinant in the feeding behaviour of a wide variety of prey species (Lima et al. 1985; Milinski 1986). According to theory, avian prey species balance predation risk with foraging needs. For instance, in a trade-off between relative safety from predators and optimal caloric gain, forest passerines tend to forage close to the protective cover of trees and bushes, whereas open-country birds stay well away from vegetation that could conceal raptors (Valone and Lima 1987; Lima 1988). In estuarine habitats, shorebirds frequent open expanses of mudflats, which allow for the timely discovery of approaching raptors. Raptors that hunt shorebirds, such as the Peregrine (*Falco peregrinus*) and the Merlin (*Falco colum-*

barius), commonly use stealth methods to take their prey by surprise (Page and Whitacre 1975; Dekker 1980, 1988; Palmer 1988; Cresswell 1996; White et al. 2002). If rising tides inundate the mudflats and force the shorebirds close to the vegetated high-tide line, most leave and fly to roosting sites in adjacent country. In some locations, shorebirds roost in habitats they do not normally frequent, such as agricultural lands (Butler 1994) or high wave-swept beaches (Buchanan 1996). On the northwest Pacific coast of North America, and possibly elsewhere in their extensive wintering range, Dunlin flocks fly out over the ocean and remain in flight for 2–4 hours until the tide turns. This behaviour was interpreted as a possible anti-predator strategy by Brennan et al. (1985) and first documented as such by Dekker (1998). Hotker (2000) saw the same phenomenon on the north coast of Germany and termed it "airborne roosting." Staying in flight over the water far offshore makes sense if roosting sites close to the high-tide mark are dangerous. Ydenberg et al (2002) found that the choice of stopover sites by migrating Western Sandpipers (*Calidris mauri*) in British Columbia, Canada, could best be explained by a hypothesis that sandpipers are more vulnerable to raptor predation on small feeding sites than on wide expanses of open mudflats.

In this article, we examine a large data set of observed hunts and kills by Peregrines to test the hypothesis that proximity to shoreline vegetation is dangerous for Dunlins, and that Dunlins tend to avoid risk when they are satiated, but are more willing to take risk when hungry. As additional evidence for that hypothesis, we also report on Dunlin kills by Merlins (*Falco columbarius*) and Northern Harriers (*Circus cyaneus*).

STUDY AREA AND METHODS

The study area was at Boundary Bay, on the southern edge of the Fraser River estuary (49°05'N, 123°00'W) in British Columbia, Canada. The bay is 16 km across and the intertidal zone is roughly 4 km wide at the lowest ebb. The tidal rhythm includes two flood tides, one usually higher than the other, per 24-hour period. During winter, the highest tides almost always occur during daylight hours and inundate all intertidal mudflats and most of the narrow strip of saltmarsh, which is covered with low vegetation. A dike protects low-lying agricultural fields inland. Boundary Bay is a major stopover for migratory waterbirds and a wintering location for circa 40,000 Dunlins and 1,000 Black-bellied Plovers (*Pluvialis squatarola*). Birds of prey are common (Butler and Campbell 1987). In winter, the bay is hunted over by at least six Peregrines and one or more Merlins (Dekker 2003).

Between early November and mid February 1994-2003, DD spent part or all of 151 days (940 hours) in the study area, walking the dike or sitting in a parked vehicle. Flocks of Dunlins were monitored for alarm behaviour such as sudden flushing. Hunting raptors were also discovered by frequently scanning the area through 8x wide-angle binoculars. Perched Peregrines and Merlins were often kept under surveillance for periods of up to two hours in the hope of seeing them hunt. We use the term "hunt" to mean a completed attack of which the outcome was known. A hunt could include one or more passes or swoops at the same Dunlin. An attack on a flock and subsequent pursuit of a single Dunlin fleeing that flock was counted as one hunt. However, if the falcon abandoned the pursuit and again attacked the same or a different flock, it was tallied as another hunt. This definition of a hunt was also used by Dekker (1980, 1988, 2003) and is the equivalent of the term "attack" as formulated by Cresswell (1996). In this paper, both terms are used interchangeably.

Field data were recorded in diary form and entered into a coded table of hunts and kills, divided over three zones. Zone #1 represented the saltmarsh shore including a 5–10 m strip of wrack and sparsely vegetated mudflat beyond the ragged marsh edge. To investigate whether Peregrine hunting success was influenced by distance from shore, we arbitrarily split the intertidal zone into zones #2 and #3. Zone #2 extended roughly 0.5 km from the saltmarsh; zone #3 lay beyond zone #2. Depending on tide height, zones #2 and #3 could consist of mudflats and/or ocean. Although a hunt or pursuit might cross over from one zone into the next, the position of the prey at the start of the attack defined the zone in which the hunt was considered to have taken place. In borderline cases, the choice amounted to a judgement call or best guess. Details on Peregrine characteristics and hunting methods, as well Dunlin behaviour when avoiding raptors, were given in Dekker (2003).

We recorded the zone (#1, #2, or #3) in which each hunt took place and the time until or since the nearest high tide. From this summary, we derived the kill rate per observation hour in relation to the time of day (intervals of one hour) and the time of the nearest high tide, based on the tide tables published for Point Atkinson, British Columbia. In a separate data analysis, RY compared the kill rate to the height of the tide (grouped into 40 cm intervals) at the time of the kill, using the algorithm available at XTide (Flater 1998). The result was tested for significance by standard regression procedures. The Peregrine hunting success rate over the shore zone (zone #1) was compared statistically to zones #2 and #3 by a G-test of independence (Sokal and Rohlf 1981).

No attempt was made to record the number of Dunlins present in the attack zone when hunts and kills took place. Large numbers of Dunlins do not necessarily translate into a high kill rate. On the contrary, singletons or isolated small flocks of prey were reported to be more vulnerable to stealth attacks by Peregrines than large or multiple flocks spaced out over a wide area (Dekker 1980, 1998; Phiollay 1982).

RESULTS

A total of 652 Peregrine hunts directed at Dunlins was observed. Of these, 94 ended in kills. Hunts and kills took place throughout daylight hours at average rates of 0.69 hunts per hour and 0.10 kills per hour. The kill rate per observation hour was quite consistent throughout the day except for declines in 08:00–09:00 and 15:00–16:00 (Figure 8.1a). The kill rate showed a strong association with the tidal cycle, rising with the flooding tide and falling with the ebbing tide

(Figure 8.1b). The kill rate peaked 1–2 hours after high tide. The only anomaly in this pattern was a marked low in kill rate beginning two hours before high tide. The rate during this one-hour interval was nearly 1/2 of the previous interval and 1/3 of the next. The kill rates also showed a strong and significant linear relation with the height of the tide, rising steadily to 0.14–0.18 kills per hour at tide heights over 4.2 m (Figure 8.1c).

The success rate of hunts differed substantially between zones and apparently as a function of the type of attack. In zone #1 all attacks were aimed at Dunlins roosting or feeding near the saltmarsh, and Peregrines always attacked using stealth, approaching very low (<1 m) over the shoreline vegetation or rushing up over the dike. Surprise could be near complete, resulting in the capture of a Dunlin the moment the flock flushed in alarm. In zone #1, Peregrines had a success rate of 44% (33 kills in 75 hunts) (Table 8.1). If a stealth hunt in this zone was not immediately successful, Peregrines rarely chased the fleeing target.

The success rate of all hunts in zone #2 was nearly the same as in zone #3 (11% vs. 10%). Combined, the success rate of stealth hunts in these two zones was 14% (21 kills in 154 hunts), significantly lower than in zone #1 ($G = 14.2$, $P < 0.001$). Most (70%) hunts in zones #2 and #3 were not based on stealth, but were open attacks on flying Dunlins or birds that had flushed well ahead of the falcon. These open attacks had a success rate of 9% (37 kills in 406 hunts). The success rate of all Peregrine attacks over zones #2 and #3 (11%) was significantly lower than all hunts over zone #1 (44%) ($G = 29.4$, $P < 0.001$).

Other raptors also attacked Dunlins. Harriers frequently attempted to approach roosting Dunlins by stealth. Of an estimated 300 harrier attacks only five were successful and all of these took place in zone #1. Probably reflecting the low risk posed by harriers, Dunlins showed minimal avoidance response. If flushed by harriers, the Dunlins returned to the same place as soon as the raptor had passed by. Merlins were far less common than harriers, but more adept hunters of Dunlins; seven of 23 stealth attacks on flocks roosting in zone #1 were successful (30%). Merlins also hunted over zones #2 and #3, using stealth as the initial strategy in 28 attacks on flocks. After the stealth approach failed, six captures were made by persistent pursuit of single Dunlins that had left the flock (21%). Attacks by Merlins and Peregrines always caused Dunlin flocks to move to another location or to begin their over-ocean flights, which generally started 1–2 hours before the rising tide inundated all mudflat habitat.

DISCUSSION

The hypothesis that raptor predation risk for small shorebirds increases with closeness to vegetation is supported in this study by three converging lines of evidence: (1) the high success rate of Peregrines hunting over the saltmarsh zone; (2) the relatively high kill rate per observation hour when Dunlins are in the saltmarsh zone; and (3) the positive correlation of kill rate with the height of the rising tide. The results of this study also lend additional support for the hypothesis that the over-ocean flocking of Dunlins during high winter tides is an anti-predator strategy (Dekker 1998; Hotker 2000). Flying far from shore, at varying altitudes depending on weather conditions, the Dunlins are safe from the most dangerous type of raptor attack: a stealth approach concealed behind vegetation.

The overall percentage of stealth hunts by Peregrines in this study (35%) is nearly equivalent to the 36% reported in 233 shorebird hunts on the coast of Scotland (Cresswell 1996). By contrast, the percentage of stealth flights in 569 shorebird hunts recorded at a large marshy lake in Alberta was 77% (Dekker 1988). The explanation for these dissimilar values is that the habitat in these two areas is quite different. At the lake, reedy shorelines create suitable cover for stealth hunts. Open attacks on flying shorebirds were uncommon at the lake except when drought had caused the shallows to recede well away from shore (Dekker 1991, 1999). By the same token, the high proportion of open hunts at Boundary Bay (62%) reflected the lack of opportunities for surprise over zones #2 and #3. Once the Dunlins were >10 m from the saltmarsh (the boundary between zones #1 and #2), the distance from shore had no significant bearing on the hunting success rate of Peregrines (i.e., kill rates for zones #2 and #3 were nearly equivalent).

While Peregrines evidently hunt and capture prey throughout the day, the fact that they killed much less often just before and much more often just after high tide cannot be related to habitat. If that were so, the kill rate per observation hour 1–2 hours before and 1–2 hours after high tide should be the same, which is clearly not the case. The most plausible reason for the observed difference is related to the behaviour of the Dunlins. Well before the cresting tide, often when the floodwaters are still >50 m from the saltmarsh, the great majority of Dunlins depart on their over-ocean flights, while others fly inland especially during or after heavy rain. The drop in kill rate during this period reflects the decreased vulnerability of the birds during over-ocean flights. (The kill rate at inland roosting sites outside the study area was not recorded.). The peak in kill rate immediately after high tide is probably due to the Dunlins' need to compensate for energy expended during 2–4 hours of over-ocean flying. Returning Dunlins are likely to be hungry, more intent on foraging and less vigilant than at other times during the tidal cycle. Consequently, the trade-off temporarily shifts toward increased risk.

A second and complementary reason for the high kill rate just after high tide is that some Peregrines, particularly the adults, spend most of the day perching and may not start hunting until flocks of Dunlins begin to congregate near the saltmarsh. Adult Peregrines are significantly more successful in the use of stealth than juveniles. Only 25% of all hunts were by adults, yet they accounted for 47% of kills (Dekker 2003).

Prey selection. Based on the examination of prey remains, a relatively high percentage of shorebirds caught by raptors are known to be juveniles (Kus et al. 1984; Whitfield 1985; Warnock 1994). The mechanics of age-related prey selection seem simple if we assume that juveniles are on the outside of flocks, either on the ground or in flight (Ydenberg and Prins 1984; Ruiz et al. 1989; Newton 1998). In this study, at least 69% of captured Dunlins were taken by Peregrines directly from the outside or tail end of flocks (Dekker 2003). Furthermore, the percentage of juveniles may be high in small, isolated flocks that render themselves vulnerable to stealth attacks during the high tide. Such risky behaviour includes: (1) late departure on over-ocean flocking flights; (2) persistence in roosting or foraging near vegetation along the high tide line; (3) early return to shore after over-ocean flocking; and (4) a switch to inland roosts or feeding sites. The proportion of juveniles in flocks in such high-risk situations, and their physical condition compared to conspecifics that flock over the ocean, might present an interesting avenue for further research.

Table 8.1. Success rates of Peregrine Falcons hunting Dunlins at Boundary Bay, British Columbia. Zone #1 represents the ocean shore-saltmarsh edge; zone #2 extends 0.5 km beyond zone #1; zone #3 begins 0.5 km from the saltmarsh. Depending on tide height, zones #2 and #3 could consist of mudflats and/or ocean.

	№ of Hunts	№ of kills	Success rate (%)
Zone #1	75	33	44
Zone #2	299	33	11
Zone #3	278	28	10
Total	652	94	14

Figure 8.1 (a, b, & c). Temporal and tidal patterns of Peregrine Falcon predation on Dunlins wintering at Boundary Bay, British Columbia, Canada.
a) Kill rate (number of captures per observation hour) in relation to the time of day. Number of kills and observation time (hours) in the successive intervals were 1/49, 11/95, 15/117, 19/128, 13/136, 15/126, 13/127, 3/110, 4/52.
b) Kill rate relative to the crest of the nearest high tide. Number of kills and observation time (hours) in the successive intervals were 2/49, 3/55, 6/62, 5/75, 4/87, 12/100, 13/106, 25/104, 18/90, 3/79, 1/69, 2/58.
c) Correlation between kill rate and height of the tide. The regression equation is: kill rate = 0.068 * tide height (m) -- 0.16. ($r = 0.86$, $F\ 1.5 = 29.8$, $P < 0.001$). Number of kills and observation time (hours) for the successive points were 0/25, 1/58/ 8/104, 16/224, 26/279, 36/199, 7/50.

a)

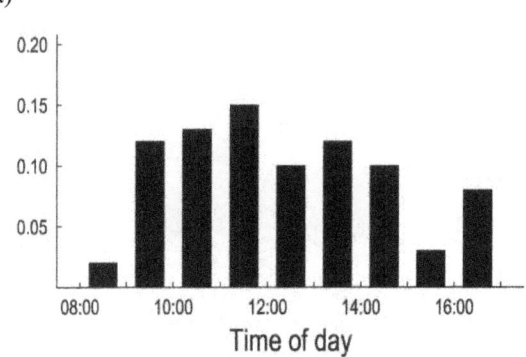

b)

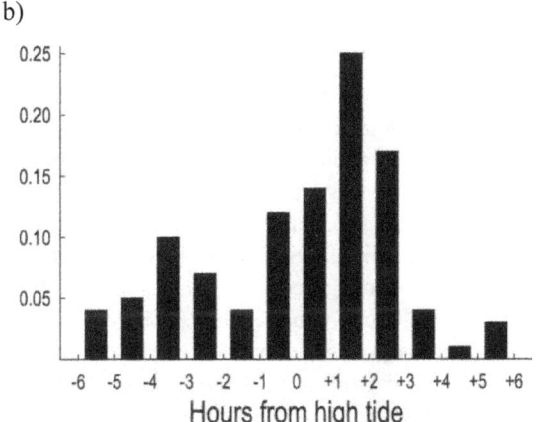

c)

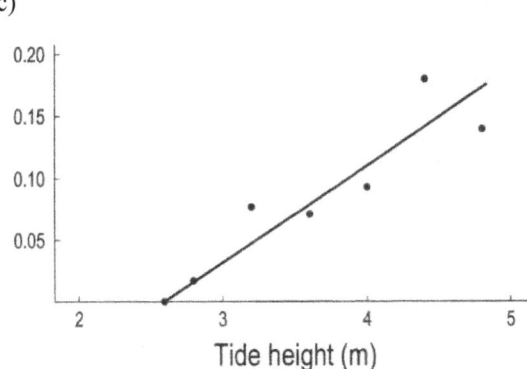

CHAPTER 9

Peregrine prey selection and eagle interference

Dekker. D. 2003. Peregrine Falcon predation on Dunlins and ducks and klepto-arasitic interference from Bald Eagles wintering at Boundary Bay, British Columbia. Journal of Raptor Research 37:91-97.

Abstract – At Boundary Bay, British Columbia, wintering Peregrine Falcons (*Falco peregrinus*) captured 94 Dunlins (*Calidris alpina*) in 652 hunts. The major hunting techniques were open attacks on flying flocks (62%) and stealth attacks on feeding or roosting flocks (35%). Success rates for these techniques were 9.1% and 23.6%. Sixty-five Dunlins were taken directly from the edge of flocks; 29 Dunlins were seized after they had split off from flocks or were flying alone. Adult Peregrines were significantly more successful than immatures (26.8% vs. 9.0%). Peregrines captured one Green-winged Teal (*Anas crecca*) and three larger ducks. The teal was carried away, but the larger ducks were pirated by Bald Eagles (*Haliaeetus leucocephalus*). Eagles often joined Peregrines that were chasing Dunlins; six eagles succeeded in either capturing the Dunlin or forcing a Peregrine to drop its just-caught prey. I postulate that Peregrines wintering at Boundary Bay abstain from capturing prey species such as ducks that are too heavy to be carried out of reach of klepto-parasitic eagles. Female Peregrines aggressively chased off other females, but tolerated males. Female Peregrines often joined males that were chasing Dunlins; 14 males were forced to surrender their prey to females. Four Peregrines pirated Dunlins from Merlins (*Falco columbarius*).

INTRODUCTION

In their nearly worldwide range, Peregrine Falcons (*Falco peregrinus*) prey on a wide variety of birds, although they may locally specialize on very few species (Brown and Amadon 1968; Palmer 1988). The size of the prey taken is partly determined by the sex of the Peregrine since females are about one-third heavier than males. This sexual size-dimorphism has spawned several hypotheses, including that it widens the range of prey that can be taken by paired falcons (Selander 1966; Cade 1982). In human altered habitats or over the ocean, both sexes of Peregrines may hunt the same prey species, namely Rock Doves (*Columba livia*) or small alcids (Beebe 1960; Ratcliffe 1993; Frank 1994;

Dekker and Bogaert 1997). But in marshes and agricultural regions, female Peregrines tend to kill larger prey than do males. For example, in central Alberta and coastal regions of British Columbia and Washington, migrant or wintering female Peregrines predominantly killed ducks, while the males took sandpipers and small passerines (Anderson and DeBruyn 1979; Dekker 1980, 1995).

In the estuaries of western North America, a common prey of wintering Peregrines is the Dunlin (*Calidris alpina*). Based on the remains of Dunlins killed by falcons, researchers in Washington and California reported that the majority are juveniles, which might be related to age-related flocking behaviour (Kus et al. 1984; Warnock 1994). This paper details the hunting methods of both sexes of Peregrines attacking flocks of Dunlins. It also reports on the duck hunting habits of these Peregrines with particular reference to klepto-parasitic interference from Bald Eagles (*Haliaeetus leucocephalus*).

STUDY AREA AND METHODS

The study area is at Boundary Bay, which is part of the Fraser River delta (49°05'N, 123°00'W) in southwestern British Columbia. The bay is 16 km across and the intertidal zone is roughly 4 km wide at the lowest ebb. The tidal mudflats are bordered by a narrow strip of saltmarsh and a dike that protects low-lying agricultural fields inland. Boundary Bay is a major stopover for migratory waterbirds and a wintering refuge for circa 20,000 ducks and 40,000 Dunlins. The only other shorebird to winter in the bay in some numbers (circa 1,000) is the Black-bellied Plover (*Pluvialis squatarola*). Bald Eagles are common year-round and increase locally to 50–150 during January-February (Dekker 1999). For a more detailed description of the Fraser delta and its avifauna, see Butler and Campbell (1987); Butler (1994); Butler and Kaiser (1995).

Between early November and mid February 1994-2003, I spent part or all of 151 days in the study area for a total of 940 hours. I walked the dike or, during rain and strong winds, sat in a parked vehicle at a vantage point from where the tide flats were visible. To study the interaction of Peregrines and their prey species, I used three principal methods: (1) flocks of ducks and shorebirds were monitored for alarm behaviour such as sudden flushing; (2) the area was frequently scanned through binoculars to spot flying Peregrines; and (3) perched Peregrines were observed for varying lengths of time in the hope of seeing them start a foraging flight.

Peregrines were classified as either adult or immature, based on dorsal colour and ventral markings. Males and females could only be separated with certainty, respectively at the lower or higher end of their size range, because recorded weights of the heaviest male Peregrines overlap with those of the lightest females, particularly between the subspecies occurring in western North America (Brown and Amadon 1968; White et al. 2002). Some falcons appeared typical of *F. p. pealei* (Beebe 1960), others of *F. p. anatum* (Palmer 1988). There is no evidence that Peregrines of arctic origin (*F. p. tundrius*) winter in the bay.

Already by the second week of February sightings of Peregrines dropped off sharply, perhaps indicative of a return to breeding sites in the region.

Data were recorded in diary form and entered into a coded table of hunts and kills. The term "hunt" means a completed attack of which the outcome was known (Dekker 1980, 1988). A hunt could include one or more passes or stoops at the same Dunlin or flock. An attack on a flock and subsequent pursuit of a single Dunlin fleeing that flock was counted as one hunt. However, if the Peregrine abandoned the pursuit and presently made another attack on the same or a different flock, it was counted as a second hunt. This definition of a hunt is equivalent to the term "attack" used by Cresswell (1996), but differs from "hunting flight" as formulated by Buchanan et al. (1986, 1996). In this paper, the terms hunt and attack are interchangeable. Whether or not a hunt resulted in a kill was not always immediately apparent, especially at distances of >1 km. If the falcon ended a hunt by flying away fast and attacking other birds, the preceding hunt was obviously unsuccessful. On the other hand, if the falcon briefly slowed down with lowered feet, or if it retrieved something from the ground or water, then headed directly to shore, it was assumed that prey had been caught. Supporting clues were pursuit by conspecifics, eagles or large gulls. However, all probable captures were deleted from the database unless they could be confirmed, for instance by feeding activity. Perched Peregrines were examined for plumage detail through a 20–60 power telescope. Separate success rates were computed for adults and immatures and for the major hunting strategies. Data collected in 1994–1998, which were presented in an earlier paper (Dekker 1998), were analysed further and added to the data obtained in 2000–2003. Data sets were compared statistically by G-test of independence (Sokal and Rohlf 1969).

RESULTS

Dunlin Hunts. In hunting Dunlins, Peregrines used two main methods: low stealth attacks on resting or feeding flocks, or open attacks on flying flocks. The stealth approach was used in 35% of hunts and resulted in 57% of total kills. The success rate of stealth hunts was significantly higher ($G = 17.53$, $P < 0.0001$) than open hunts (Table 9. 1). Low stealth attacks were launched from perches or flapping flight, rarely by a stoop from high soaring flight. The Dunlins usually flushed at the very last moment and the Peregrine might immediately succeed in seizing its prey. Often, one or more Dunlins dodged the falcon by dropping back onto the mud or water. Most took off again at once into the opposite direction after the Peregrine overshot its mark. Others failed to rise and appeared to have been hit. These prey, either crippled or dead, were retrieved from the water in a return pass by the falcon.

In 62% of hunts the falcons attacked flocks of Dunlins flying over the mudflats or ocean. At the approach of a Peregrine, the Dunlins formed dense, globular flocks. (For a detailed description of Dunlin evasive flocking behaviour, see Buchanan et al. 1988). If the flock was actually singled out for attack, Peregrines used a variety of methods. They commonly flew back and forth 10–100 m above the Dunlins, manoeuvring for position. Eventually, they stooped perpendicularly with rigid wings, either held open or tucked in. From a distance, depending on the angle of observation, it might seem as if some falcons plunged right through the flock, but this could not be verified. Instead, most falcons skirted the outside edge of the flock and aimed their attack at its bottom (Figure 9.1a). However, some falcons stooped at the top of the flock, which reacted by caving in and flattening out low over the water (Figure 9.1b). To evade the falcon, one or more Dunlins might splash down and take off again at once. Others might be hit and crippled, unable to rise. In the final stage of the attack the falcon could: (1) seize its prey at once or double back and retrieve a downed Dunlin from the water; (2) pursue a single Dunlin that split away from the flock; (3) regain altitude and manoeuvre for another vertical stoop at the same flock; (4) attack a different flock some distance away; or (5) end the hunting sequence by flying away.

If approached by low-flying falcons (<1 m), flocks of Dunlins that flushed ahead might stay low as well, particularly during stormy weather. If attacked and overtaken, Dunlins in the trailing end of the fleeing flock could be seized at once (Figure 9.1c). Single birds that dodged by dropping into the water and taking off again might be pursued and swooped at with varying degrees of persistence. Twenty-seven Dunlins were captured after individual pursuit including 3–10 passes. Two lone Dunlins, flying at 20–25 m, were seized by immature male Peregrines that stooped nearly perpendicularly from a height of >100 m. In three separate but unsuccessful instances immature Peregrines persisted in follow-chasing Dunlins >100 m high and over distances of >1 km. After dodging 20–40 swoops, all three Dunlins fled inland and eventually plunged into bushes.

The main hunting methods of stealth and open attack were used by all Peregrines, although open attacks on flying flocks with multiple stoops appeared to be more typical of males than females. Immature males could be very persistent but quite unsuccessful. For instance, on 21 January 2003, an immature male failed to make a capture over a period of 2.5 hours during which he launched three series of attacks over the low tide line, including a total of about 30 stoops at flocks. On 24 and 25 January, an immature male made 22 and 33 consecutive stoops at flying flocks respectively, all without success. By contrast, on 11 November 2002, an adult male captured two Dunlins 20 minutes apart, each requiring only one stealth attack on roosting flocks. (This male was robbed of his first kill by an adult female Peregrine). Of the total 652 hunts directed at Dunlins, 25% were made by adult Peregrines, yet they accounted for 47% of the kills (Table 9.2). The success rate of adults was 26.8%, significantly higher than the 9.0% rate of immatures ($G = 19.92$, $P < 0.0001$).

Unless they were hunting, adult males appeared to spend little time in the study area. On the other hand, adult females could be found perching for hours on driftwood snags in the salt marsh or on inland fields. Evidently territorial, they attacked and chased off other females, but tolerated males and often joined them in attacks on flocks of Dunlins. Frequently, both sexes pursued the same prey. However, if the male made the capture, he was chased and robbed by the female. I saw 14 males release and lose their catch. Some chases covered long distances with undetermined results. Probably to avoid piracy by conspecifics, male Peregrines, upon capturing a Dunlin, flew to forested headlands across the bay or carried their prey far inland. Others circled to a high altitude and consumed their prey on the wing. Peregrines, either male or female, also robbed four Merlins (*Falco columbarius*) of their just-caught prey.

Dunlins were hunted throughout the day; 46 kills were made prior to 12:00 and 48 after 12:00. The respective values for kills per hour of observation (1/8.1 and 1/12.4) in morning or afternoon were not significantly different ($G = 2.85$, $P > 0.05$). Peregrines often resumed hunting immediately after eating a Dunlin, which took 10–20 minutes.

Duck Hunts. The Peregrines rarely hunted ducks in the coastal study area. Only four kills were seen. A Green-winged Teal (*Anas crecca*) that was flushed and briefly pursued by an adult male was taken after it dropped into a shallow pool in the saltmarsh. The male carried his prey out of sight low over the vegetation. Female Peregrines pursued and captured three ducks larger than teal, presumably Northern Pintails (*Anas acuta*), which were seized on the tide flats after the ducks dodged the approaching falcon by plunging into shallow water.

Klepto-parasites. All three large ducks captured by Peregrines in the study area were forfeited without a struggle to Bald Eagles. On four occasions, duck-hunting Peregrines were followed by one or more Bald Eagles, which attempted to seize ducks that dodged the falcons by ditching into the water. Eagles also commonly joined and interfered with Peregrines that were hunting Dunlins. One or more eagles immediately started in pursuit of most falcons (50–60%) that had just caught Dunlins. The Peregrines usually managed to stay ahead, but three falcons, closely harassed by 2–4 eagles, released their Dunlin. Two of these falling prey items were retrieved in mid-air by eagles. In at least three cases, eagles pounced on Dunlins that had ditched into the water or dropped into grassy vegetation to dodge the falcon.

The following incident is a typical example of intra- and interspecific competition between falcons and eagles. In the afternoon of 16 January 2003, a male Peregrine stooped at a flock of Dunlins and pursued a single bird that split off and dodged repeated passes. The male was quickly joined by a female and four eagles, all of them alternately swooping at the Dunlin flying erratically 1–10 m high over the water. After a combined total of 10–12 passes, the Dunlin was caught by the female Peregrine, which was immediately set upon by the eagles. The falcon carried her prey far inland. Prey-carrying Peregrines were frequently but vainly chased by Glaucus-winged Gulls (*Larus glaucescens*). In addition, gulls often joined Peregrines that were pursuing lone Dunlins, and on two occasions gulls pounced on Dunlins that had ditched into the water to escape a swooping Peregrine.

DISCUSSION

The stealth tactics used by the falcons in this study were similar to those reported for migrating Peregrines hunting shorebirds in central Alberta, although, due to more favourable weather conditions, the Alberta migrant falcons launched their stealth attacks more often from high soaring flight (Dekker 1980, 1988, 1999). The percentage of stealth hunts in the two studies is however quite different (35% at Boundary Bay versus. 77% in Alberta). The explanation can be found in differences in the respective habitats. The Alberta falcons hunted over reed-studded shallows of a large inland lake, which afforded ample opportunities for a concealed approach. By contrast, the relatively low use of stealth on the coast of British Columbia reflects the lack of opportunity for effective surprise over the tide flats. The percentage of stealth hunts in this study is nearly equivalent to the 36% reported from the east coast of Scotland (Cresswell 1996).

It is possible that stealth hunts, especially successful ones by adult Peregrines, are under-recorded in this study, resulting in a bias in favour of open hunts. Successful stealth attacks on flocks of Dunlins roosting near the saltmarsh easily escape detection by humans. Stealth attacks are often initiated from distances of >1 km and the terminal portion takes place very low over the ground or water. The light-coloured plumage of the adult falcon blends into the background of water and renders it more difficult to detect than a dark immature Peregrine. Screened by the massive flushing of Dunlins, the attacking Peregrine might carry its prey away before the observer is aware of what happened. Adult falcons are experts in the use of stealth and especially the adult male tends to keep a low profile, perhaps to avoid attracting the attention of klepto-parasites including female Peregrines.

In contrast to stealth attacks, open hunts are readily seen because the observer is alerted by the rising of dense flocks of Dunlins. However, it takes skill and experience to discover the falcon that may still be far off and either fly very high or low. Furthermore, during a prolonged sequence of attacks, it is difficult to keep accurate score of the exact number of hunts. Stoops on flocks can follow each other in rapid succession and the observer cannot always be certain that all were aimed at the same or different flocks. To avoid this problem, Buchanan et al. (1986) used the term hunting flight, which is defined as "a perch to perch flight involving one or more capture attempts." Although his sample of kills was very small ($n = 7$), Buchanan et al. (1986) calculated the success rate of hunting flights as 47% and that of individual capture attempts as 14.6%.

Any study of the hunting habits of the Peregrine Falcon involves a certain amount of subjective interpretation, which might in part explain the extreme variance between studies (White et al. 2002). In my appraisal of open hunts, I kept score of separate attacks in a conservative way, concentrating on the outcome. How the kill was made seemed of more importance than how often the falcon missed. Many hunts that fail may not be all that serious. This possibility was pointed out by Treleaven (1980), who coined the terms high-intensity and low-intensity hunting.

Based on the examination of prey remains, a high proportion of shorebirds killed by raptors were reported to be juveniles (Whitfield 1985; Warnock 1994). Kus et al. (1984) suggested that this might be due to age-related differences in Dunlin flocking behaviour. I postulate that the mechanics of prey selection appear to be very simple. The majority (69%) of Dunlins captured during this study were taken from the outside or the tail end of flocks (Figure 9.1). Flocking is a well-known predator avoidance behaviour of open-country birds. Their drawing together into dense aerial formations is the result of each bird's instinctive desire to find safety in the centre of the flock (Tinbergen 1951). I speculate that adults are more proficient at this than juveniles, which are left on the outside of the flock. Several studies have reported age-related segregation of birds in roosting and feeding flocks (Newton 1998). If Peregrines selectively remove juvenile Dunlins from a wintering population because they are easier to capture than adults, then the proportion of vulnerable juveniles in that population declines over the course of the winter. Furthermore, juveniles that are attacked on a daily basis and manage to survive for several months might learn to become more vigilant and avoid capture. These hypotheses seem to be supported by the data. The respective hunting success rates of Peregrines for November and January declined from 18.6% ($n = 134$) to 12.4% ($n = 404$), although the difference is not significant ($G = 2.30$, $P > 0.05$).

The klepto-parasitic habits of Bald Eagles are well-known, particularly at the expense of Peregrines that hunt ducks or seabirds (Anderson and DeBruyn 1979; Dekker 1995; Dekker and Bogaert 1997). Bald Eagles also klepto-parasitize Merlins. During this study, Merlins were seen to capture 14 Dunlins, two of which were pirated by pursuing eagles. Buchanan (1988) reported similar instances and he postulated that the threat of losing prey to larger raptors resulted in Merlins engaging in hunting flights of shorter duration if potential kleptoparasites were present. Gyrfalcons (*Falco rusticolus*) wintering in Alberta were also often robbed of ducks by eagles and avoided hunting at localities where Bald Eagles sat on prominent perches (Dekker and Court 2003).

In conclusion, I suggest that the ubiquitous presence of eagles at Boundary Bay discourages wintering Peregrines from hunting ducks on the tide flats. The possibility exists that Peregrines may hunt ducks more often at night or at inland localities where eagle presence is lower than at the coast. Adult female Peregrines, which had perched on the coast for much of the day, flew inland at dusk when wigeon leave the coast to feed on inland meadows (Dekker 1999). During November, when eagles are far less numerous at Boundary Bay than in January, several immature Peregrines persistently hunted ducks in the study area (Dekker 1998; and unpublished data). As the winter progresses however, I surmise that these juveniles eventually stop hunting ducks on the tide flats and instead concentrate on prey such as the Dunlin that can easily be transported over long distances, out of reach of klepto-parasitic eagles.

Table 9.1. Success rates of various strategies employed by Peregrine Falcons hunting Dunlins at Boundary Bay, British Columbia.

Method	Hunts	Kills	% Success
Stealth hunts	229	54	23.6
Open hunts	406	37	9.1
Unknown method	17	3	17.6
Totals	652	94	14.4

Table 9.2. Success rates of Peregrine Falcons hunting Dunlins in British Columbia, and comparison with other major studies of Peregrines hunting shorebirds.

	Hunts	Kills	% Success
This study (Seacoast, winter)			
Adults	164	44	26.8
Immatures	399	36	9.0
Unidentified	89	14	15.7
Totals	652	94	14.4
Scotland* (Seacoast, winter)	233	25	10.7
Alberta** (Prairie Lake, migration periods)	569	50	7.2

Sources: *Cresswell 1996. **Dekker 1988

Table 9.3. A comparison of the prey selection of Peregrines hunting at Boundary Bay during the winter months and fall.

	Total hunts	Kills of Dunlin	Kills of ducks
This study	652	94	4
Additional winter data	167	26	1
Fall data	70	4	9

Figure 9.1. Peregrine modes of attack on aerial flocks of Dunlins. The prey is either taken from the bottom, the outside edge, or the tail end of the flock.
(A). Falcon stooping alongside of the flock and aiming for the bottom.
(B). Falcon stooping directly at the flock, which caves in and flattens out low over the water.
(C). Falcon pursuing just-flushing or low-flying flock.

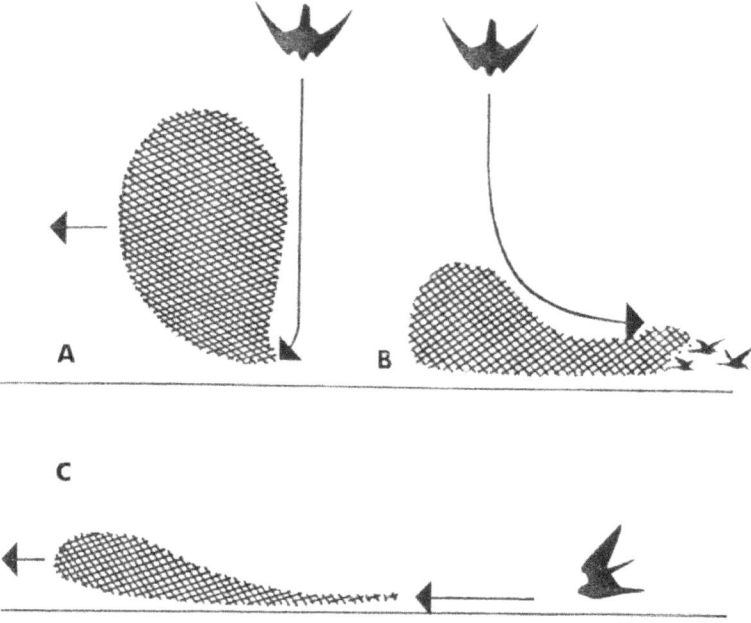

Addendum. In January and February of 2006 and 2008, I spent an additional 30 days at Boundary Bay and saw a total of 167 hunts by Peregrines, which resulted in 26 captures of Dunlins and one duck. By comparison, during 21 days in November 2003 and October 2006, my tally of hunts was 70 including four captures of Dunlins and nine ducks (Table 9.3). The difference between these sets of data was highly significant ($P < 0.001$), supporting the hypothesis that Peregrines hunt ducks as well as Dunlins in the fall when eagle presence is low, and Peregrines stop hunting ducks in winter when eagles are ubiquitous at Boundary Bay.

CHAPTER 10

Cooperative hunting by parent Peregrines and their fledglings

Dekker, D. and R. Taylor. 2005. A change in foraging success and cooperative hunting by a breeding pair of Peregrine Falcons and their fledglings. The Journal of Raptor Research 39:386–395.

Abstract – The foraging habits of one pair of Peregrine Falcons (*Falco peregrinus*) nesting on a powerplant in central Alberta were studied over seven consecutive breeding seasons (1998–2004). We observed 386 attacks that resulted in 117 captures of prey. The success rate increased from 21.9% in the first year to 39.1% in the seventh year. The majority of hunts (76.7%) were initiated from soaring, and the Peregrines commonly used the hot air above the plant's smoke stacks to gain height. The success rates of hunts launched from soaring or from perches were not significantly different (28.7% vs. 35.6%). Tandem hunts by the pair ($n = 100$) were more successful than solo hunts (39.0% vs. 27.3%). The main prey species were Franklin's Gulls (*Larus pipixcan*) and small passerines, which made up 53.0% and 27.4% of kills with respective capture rates of 42.5% and 24.1%. The success rates of the male and the female Peregrine were not significantly different, but there was a notable difference in the prey taxa taken by each gender. The male captured 84.4% of the passerines but only 22.9% of the gulls. However, after the first year, while foraging for his mate or the young, the male switched from mainly passerines to gulls. All gulls captured by the male were surrendered to the female or the fledged juveniles. In 23.3% of hunts, one or both parents were accompanied by fledglings. The male participated in 76.7% of all parent-fledgling hunts ($n = 90$), the female in 10.0%, and the remainder by both parents. Aerial prey transfers from adults to flying young were feet-to-feet or through aerial drops. The hypothesis that parent Peregrines make live-drops of prey to their fledged young to teach them how to hunt is refuted by observations that live-drops of just-caught prey are also made by the adult male to his mate. However, the hypothesis that Peregrines assist their young in capturing prey is supported by anecdotal evidence.

INTRODUCTION

Cooperative or tandem hunting by mated pairs of Peregrine Falcons (*Falco peregrinus*), in which both partners attack the same prey simultaneously, has been reported from many locations across the species' nearly worldwide range (Cade 1960; Bird and Aubry 1982; Thiollay 1988; Frank 1994; Treleaven 1998; Jenkins 2000). The success rate of tandem hunts has been reported to be higher than hunts by single Peregrines, but whether or not tandem hunting by mated pairs actually involves a degree of coordination between the two individuals is unclear. On their breeding range, Peregrine Falcons are also known to hunt together with their fledglings although data are limited (Brown and Amadon 1968; Palmer 1988; White et al. 2002). The notion that juvenile Peregrines need to be taught hunting skills by their parents is contradicted by the fact that captive-reared Peregrines, deprived of parental instruction, begin pursuing and capturing prey at about the same age as young falcons at natural nest sites (Cade and Burnham 2003). Apparently, the Peregrine does not need to be taught how to chase and kill prey (Cade 1982; Sherrod 1983). However, Newton (1979) and Ratcliffe (1993) state that more critical observation is needed during the period in which fledglings achieve independence. This paper presents empirical data on the foraging behaviour of a mated pair of Peregrines hunting solo or in tandem over seven consecutive breeding seasons. Additionally, we present anecdotal observations on parent/fledgling interaction.

STUDY AREA AND METHODS

The study area is in central Alberta, Canada, at 53°N. In this largely agricultural region, Peregrines nested commonly on cliff banks along rivers and creeks until they became extirpated (Dekker 1967; Fyfe 1976). A government program of releasing captive-reared Peregrines in central Alberta began in 1975 and led to the establishment of breeding pairs on city buildings by 1981 (Holroyd and Banasch 1990). By the mid 1990s, Peregrines began using nest boxes on high industrial structures in rural regions. The breeding site selected for this study is a powerplant on the north shore of Wabamun Lake, which is roughly 8 x 20 km. The nest box was built by the Alberta Falconry Association in 1993 and put in place by TransAlta Utilities (Wabamun, Alberta) on a catwalk just below the top of a 91 m smoke stack. Lower down, the flat roof of the main building functions as a landing pad for the newly fledged young. Electricity pylons in the vicinity provide high perches and plucking posts. Settling ponds and open ground adjacent to the plant are used as loafing areas by the fledglings and by the adults to bring down large prey such as gulls and ducks. The landscape around the Wabamun plant is mainly wooded with small marshes and an extensive area (5–10 km²) of excavated and reclaimed terrain to the north. The lake receives light recreational use and some shoreline sections are developed with cottages and small marinas.

The nest box is in plain view from a public road that runs between the lake and the plant. Depending on weather conditions, we watched from different vantage points 0.1--2 km away, and we used 8x wide-angle binoculars and 20–60x telescopes. Observation periods, lasting 3–6 hours each, usually between mid-afternoon and sundown, were spaced arbitrarily and increased during the period when the fledglings were on the wing. Over seven years, 1998--2004, the number of observation days was 196: 15 in June, 58 in July, 101 in August, 21 in September, and one in October. The only two days of observation in 1997 were added to the 1998 total. The median date when the first juvenile (always a male) fledged was 16 July (13--25 July), and the number of young was 3–4 per season with a mean of 3.6 ($n = 11$ males and 14 females). The breeding pair consisted of the same individuals for the duration of the study. In 1998, the black-banded female was three years old, based on the date of her banding as a nestling in the only wild nests known in central Alberta. The red-banded male originated from a captive breeding program of *anatum* stock (Gordon Court and Bruce Treichel, Alberta Environment, personal communication.).

Prey species upon which Peregrines are known to feed, such as waterbirds and small passerines, are common in the study area. With a mass of 220--335 gram (Dunning 1984), the Franklin's Gull (*Larus pipixcan*) is close to the mass limit that male Peregrines can or want to carry over long distances. Although Franklin's Gulls are locally scarce in early summer, large migrating flocks begin arriving in mid July. Rock Doves (*Columba livia*), which are the dominant prey for Peregrines in human altered environments (Ratcliffe 1993), were resident at the plant site but only in small numbers (<20).

The terms hunt and attack are used interchangeably and represent one completed attempt at capturing prey including one or more stoops or repeat passes at that same prey of which the outcome was known. Tandem hunts were simultaneous attacks by both adults on a flock of prey or a prey individual. Group hunts by a combination of parents and juveniles were tallied as adult hunts, and their kills were considered to have been made by the adults, although in a few cases it was the juvenile that actually seized (or was allowed to seize) the prey. Hunts by juveniles alone were not tallied. Details of hunts and kills were entered into field diaries and coded tabulations. Observer bias in comparing results between years, we believe, was not a factor as DD recorded observations over the entire 7-year study period. RT was associate observer during the last three years of the project. Data sets were compared for significance by Chi-square Tests of Independence with a Williams' correction as described in Sokal and Rohlf (1981).

RESULTS

Hunting methods and prey species. We observed 386 hunts by the adult Peregrines. Nearly all (99.5%) were directed at airborne prey and consisted of two primary methods: attacks launched from soaring flight (76.7%) or from a perch (23.3%) (Table 10.1). The falcons typically began a soaring sequence by circling up over the plant and using the hot air of the three smoke stacks to rapidly gain height. Drifting downwind, the soaring falcons reached altitudes estimated at >1 km. By flapping their wings or gliding, they cruised upwind. Attacks on prey flying lower than the falcon were made by deep stoops with wings partly or fully flexed. Some of these attacks began with a burst of wing beats and ended in a stoop, which could either be near vertical or oblique. If the prey evaded the initial stoop, the falcon might follow-up with additional stoops or passes.

Still-hunting attacks were launched from high perches such as the catwalk railing near the top of the 91 m stacks. With rapid wing beats, these falcons headed for targets some distance away (>100 m). After unsuccessful or aborted attacks, the Peregrines commonly returned to the plant, either to perch or to regain altitude by soaring. Some hunting sequences lasted 3–4 hours before a prey was captured. On other days, the falcons were seen to catch three preys in less than one hour.

The success rate of all hunts ($n = 386$) by the adult Peregrines, either attacking prey solo or in tandem, was 30.3% (Table 10.1). Perch hunts were not significantly ($G = 0.77$, $P = 0.379$) more successful than soar hunts (35.5% vs. 28.7%). The success rate of the female was 37.1%, not significantly ($G = 2.37$, $P = 0.124$) different from that of the male (24.1%), but there was a notable difference in the taxa of the prey taken by the sexes (Table 10.2). The male caught 84.4% of 32 small passerines, but only 22.9% of 62 gulls. Nearly all gulls seized by the male were released to his mate or a fledgling. If neither the female nor any of the fledglings were nearby, the male, upon seizing a gull, brought it down steeply and left it on open ground or on the roof of the plant ($n = 5$). The female carried gulls with apparent ease over distances exceeding 1 km.

The majority (84.9%) of gull kills observed at close range or found as prey remains ($n = 73$) were juveniles. Juvenile Franklin's Gulls were often seized in mid air at the falcon's first stoop. By contrast, adult gulls typically evaded the stoop by rising steeply. Some dodged two or more additional passes and were eventually left alone. Others descended and plunged into water. In seven instances, the female Peregrine repeatedly swooped at swimming gulls. Three were retrieved from the water and carried off. Some white birds, assumed to be gulls, that evaded 20 or more swoops far out over the lake probably were Common Terns (*Sterna hirundo*). Two terns were carried to the plant, but their capture had not been observed. Ring-billed Gulls (*Larus dela-warensis*) were sometimes forced down, but no evidence of kills was found (until after the study had been terminated).

Rock Doves were seldom attacked. On two occasions, the female stooped unsuccessfully at free-flying flocks of doves. Both adults made opportunistic passes at doves that flushed from plant ledges below them, but the only successful dove hunts ($n = 2$) were initiated by fledglings. In at least one of these hunts, the capture was made by the adult female who joined the attack.

During the sixth year of study, we gained the impression that the falcons had become more successful (in particular at capturing gulls) than during the preceding years. This impression was strengthened in year #7 and confirmed by sub-

*The hunting success rate of a pair of adult Peregrines nesting on an electricity generating plant in central Alberta rose from 22% in their first year to 39% in their seventh season of residence. During the next three breeding seasons the rate reached 46%. The pair largely specialized on catching first-year Franklin's Gulls (*Larus pipixcan*). (Photo: Gordon Court).*

sequent data analysis. The overall success rate of the adults increased from 21.3% in the first year to 39.1% in the seventh year (Figure 10.1).

Tandem Hunts. Each year from mid June onwards, both parents often began their hunting flights by soaring together. The male gained height quicker and always circled higher than the female. Stooping alternately, they attacked 68 preys in tandem (Table 10.1). If no prey had been sighted for some time, the pair separated or, following each other, headed off into the distance in high sailing or flapping flight. Tandem attacks were also launched from high perches on the plant. In both methods, either falcon could take the lead and the male flew higher than the female. While the female approached the prey directly, the male typically attacked from above in a near vertical stoop. A high percentage (60--75%) of tandem hunts became lost from view before we could either see the target or the result. Tandem hunts of which the outcome was known ($n = 100$) represented 25.9% of all hunts and their success rate was higher than in 286 solo hunts (39.0% vs. 27.3%), but the difference was not significant ($G = 2.39$, $P = 0.121$). The success rate of 32 tandem attacks launched from a perch was 50.0%, compared to 33.8% for tandem attacks from soaring.

Tandem hunts resulted in 39 kills (Table 10.1). Of these, 26 captures of prey were observed in detail: seven were seized by the female, 19 by the male. Prey caught by the male in tandem attacks were surrendered at once to the female, either by feet-to-feet transfer ($n = 5$) or released in the air ($n = 14$). Thirteen of these aerial drops were gulls, which resumed flight upon release. Nine were subsequently seized by the female; four gulls evaded her passes and were let go or escaped by splashing down into water.

Hunts by groups of adults and fledglings. In 23.3% of hunts ($n = 386$) one or both parents were accompanied and often harassed by juveniles (Table 10.3). Some fledglings closely followed the adults prior to the start of a hunt, others approached hurriedly from a distance to join a hunt in progress. A significant majority of group hunts ($G = 37.96$, $P = 0.001$) were led by the adult male, which made 71.0% of the 31 kills (Table 10.3). Prey caught by the parents were transferred, usually at once, to the approaching juveniles, either feet-to-feet ($n = 15$) or through aerial drops ($n = 7$). At least three small passerines that were dropped were still alive and resumed flight. They were recaptured by a juvenile or by the adult and released again. Three live-drops were gulls. In addition, four of 11 gulls caught in adult-juvenile hunts were seized in flight by the juveniles chaperoned by adults.

After successful group hunts, one or more juveniles fed on the kill. If the feeding site was in an open area away from the plant, the adult female typically perched on a pole nearby. At the approach of people, she vocalized and swooped overhead. She also stooped at crows or buteo hawks to drive them away. After the fledglings were satiated and left, the adult female fed on the remains of the kill ($n = 5$).

DISCUSSION

Hunting success. The hunting success of adult Peregrines on breeding territory is generally higher than that of migrating or wintering Peregrines (Dekker 1980; Roalkwam 1985). Jenkins (2000) compared data presented in eight publications on breeding Peregrines and found an extreme variation (9.3%--84.1%) in hunting success, but much less variation in hunting methods. The majority (58%–75%) of attacks summarized in his review were launched from a perch. By contrast, the percentage of perch hunts was only 23% at Wabamun. The "still-hunting strategy" of perch hunts reduces the energy cost of foraging. However, high soaring flight is also energy-efficient and widens the radius of the attack zone (Enderson and Craig 1997). Soaring and sailing flight have been reported as a common hunting method in many regions, but not to such a high degree as documented in this study. The second highest use was recorded in Africa, where breeding pairs launched 30%–42% of hunts from flight (Thiollay 1988; Jenkins 2000). The use of chimney exhaust to facilitate soaring flight by the Wabamun Peregrines was described in an earlier publication (Dekker 1999), but this phenomenon has to our knowledge not been reported elsewhere.

It is noteworthy that the falcons in this study attacked nearly all of their prey in flight, often at great altitudes. This is in sharp contrast to migrating or wintering Peregrines, which commonly use surprise methods to attack prey on the ground or in shallow water (Dekker 1980, 2003; Cresswell 1996). However, there is also an element of surprise involved if a soaring Peregrine stoops from a great height at aerial prey far below, such as small passerines flying just beyond a line of trees. Surprise is also a likely factor in tandem hunts in which the male stoops perpendicularly from high above at a prey whose attention is focussed on the female falcon attacking the same prey lower down.

A key factor in the hunting success of Peregrines is the vulnerability of individual prey, which is difficult to assess for the human observer. It is well known that predation risk for land birds increases over water (Herbert and Herbert 1965). Conversely, waterbirds become vulnerable over land (Hunt et al. 1975; Dekker 1980). Mature prey on home territory should be harder to catch for a raptor than juvenile prey passing over unfamiliar terrain. In central Alberta, the migrations of juvenile prey species roughly coincide with the period when fledgling Peregrines are on the wing and when the parental task of provisioning them is most demanding. In this study, the Peregrines were very successful at capturing juvenile Franklin's Gulls. For instance, on seven days between 17 July and 9 August 2003, when numerous flocks of gulls were passing by the powerplant, the adult female, hunting solo, captured each of seven juvenile gulls in her first attack of the afternoon and each requiring only one stoop. While it does not seem surprising that bird-hunting falcons should become better at what they do as they become older and gain experience, data in support of that notion have, to our knowledge, not been published before. An explanation for the year-to-year increase in the success rate of the Wabamun pair is that these falcons became specialists on juvenile Franklin's Gulls, which differ from adults by their dusky colour and absence of black on the head. Juvenile targets were probably preselected by the falcons. A similar pre-selection hypothesis was advanced for the high success rate (73%–93%) of "Red Baron," an adult male Peregrine hunting over coastal marshes in New Jersey (Cade and Burnham 2003: 333).

In this study, the prey component changed significantly after the first year (Tables 10.4 and 10.5). In year #1, gulls made up only 8.7% of kills as compared to 63.8% in years #2–7 ($G = 10.85, P = 0.001$). The proportion of small passerines changed significantly from 60.9% in year #1 to 19.1% in years #2–7 ($G = 7.01, P = 0.008$). Overall, the capture success on gulls (of both age groups) is significantly higher ($G = 5.35, P = 0.021$) than on small passerines (42.5% vs. 24.1%). Success rates on juvenile gulls can be expected to be even higher than for adult gulls, but we did not keep separate records. As reported in the section on results, our sample of gulls captured and remains found ($n = 73$) included 85% juveniles. A switch from small prey to large prey should have survival value and boost the provisioning rate for Peregrine young.

Coincident with the observed prey switch, we recorded a major change in foraging participation between the sexes after the first year (Table 10.6). The male's solo hunts in year #1 were 78.7% of total hunts ($n = 108$), significantly greater ($G = 8.08, P = 0.004$) than the 47.1% of hunts in years #2–7 ($n = 278$). By contrast, the female's share in year #1 was 2.7% and rose significantly to 24.1% in years #2–7 ($G = 23.51, P = 0.001$). A possible explanation is that the female lacked skill and experience in her first year on territory. As she became older and gained expertise, she began to play a more active role in foraging. There was no significant change ($G = 2.70, P = 0.1$) in the proportion of tandem hunts between years #1 and #2–7 (Table 10.6). Our findings that the main prey of the male consisted of passerines whereas the female's main prey was gulls (Table 10.2), lends further support to the hypothesis that the reversed sexual size dimorphism in raptors such as the Peregrine allows them to exploit a wider prey base and reduces competition between the sexes (Selander 1966).

Cooperative hunting. Are Peregrines that attack the same prey simultaneously with their mates or fledglings actually cooperating? Or are the individuals simply reacting at the same time to the stimulus of sighting prey? True cooperative hunting differs fundamentally from other forms of group predation, such as pseudo-cooperative hunting (Ellis et al. (1993). In true cooperative hunting, the group consists of at least two members that are a stable social unit, and their cooperation should benefit the group rather than just the individual. An example of true cooperative foraging is described by Bednarz (1988) for the Harris' Hawk (*Parabuteo unicinctus*). Ellis et al. (1993) conclude that cooperative hunting is the most efficient strategy for capturing prey in many situations, and that each form of social foraging should have evolved as an adaptive advantage enhancing the fitness of all individuals in the group. In the Aplomado Falcon (*Falco femoralis*) pair hunting is more than twice as productive as solo hunting when the prey are birds (Hector 1986). Similarly, the success rate of pair hunting in Peregrines may depend on the species of prey hunted (Jenkins 2000).

In this study, tandem hunts by the mated pair of Peregrines were more successful than solo hunts (39.0 vs. 27.3%), although the difference was not statistically significant. Moreover, the individual success rate of the two partners in tandem hunts is actually halved. So what, if any, is the fitness value of tandem hunting for this pair of breeding peregrines? The answer is that any action by the male that benefits his mate or their progeny should be of adaptive advantage.

At Wabamun the adult male was not seen to eat the gulls he killed, although on two occasions he fed on gull remains abandoned by fledglings. The fact that he mainly hunted such prey in the company of his mate or fledglings suggests that his role was a cooperative one. This view is supported by anecdotes such as the following. By mid September 1999, after the fledglings had dispersed, the adults continued to hunt together. One evening, shortly after both had soared up over the plant, the male caught a small passerine and landed on a pylon to consume his prey. Instead of interfering, the female perched on the next pylon and waited for her mate to finish his meal. Presently, both soared up again and eventually headed out high over the lake, stooping in tandem at gulls. Like many similar observations, this incident suggests that the male's role was indeed cooperative and intended to assist his mate in her foraging.

The cooperative nature of tandem hunts by adults and fledglings seems less clear, for it is apparent that the juvenile is primarily intent on klepto-parasitising the adult. If there is cooperation, it is one-sided and benefits only the juveniles. Nevertheless, the evolutionary value of adult/fledgling combinations is undeniable as a well-fed progeny enhances the future of the family genes. In support of the hypothesis that the participation of adults in joined hunts with their fledglings is indeed cooperative and represents parental care, we present a number of anecdotal observations in the appendix.

Live-drops of prey. Cade (1982) and Sherrod (1983) characterized the behaviour of paired male Peregrines in social interactions with their bigger mate as submissive and even fearful. Females take food from males in an aggressive manner, not only during the breeding season but also on migration or at the wintering grounds. Unmated females routinely force males to drop just-caught prey (Dekker 1980, 2003). Tandem hunting by unmated Peregrines is common and driven by competition rather than cooperation. Two or more (up to six) migrating or wintering Peregrines were seen in joint pursuit of the same prey in Alberta and British Columbia. Male Peregrines were forced to make live-drops of prey not only to female conspecifics, but also to other klepto-parasitic raptors, such as Prairie Falcons (*Falco mexicanus*) and Bald Eagles (*Haliaeetus leucocephalus*) (Dekker 1995, 1998).

I concur with Sherrod (1983) that the release of just-caught prey by the male Peregrine at the approach of an aggressive female may simply be triggered by the impulse to avoid close contact. The importance of a timely release was demonstrated on 21 August 2003 when the Wabamun male, carrying a just-caught prey and flying 20–25 m high, was met by a screaming juvenile female in typical begging flight. Apparently, the male was just too late in releasing his prey, for he was seized by the juvenile. Locked together by their feet, both birds tumbled about 15 m before separating again, while the prey item fell into the bushes below and was lost.

Based on a thorough review of the literature and on his own studies, Sherrod (1983) observed that the behaviours of parents and fledglings complemented each other. While the youngster only wants food, the adult appears to be very willing to supply that food. Some live-drops are no doubt the result of the aggressive approach of the begging falcon. However, as argued by Newton (1979), other live-drops are clearly intentional since parent Peregrines will recapture the live-dropped prey and release it again after the fledgling fails to catch it the first time. Such repeat drops were seen in this and other studies.

Conclusion. Some of the aerial drops of just-caught and still alive prey from adult to fledgling (described in the appendix) lend support to the hypothesis that the parents were teaching their young by example. Although young falcons do not need to be taught to hunt, such extended parental care might well give them a certain survival advantage after leaving the nest site. The observation that live-drops were also made from adult to adult, both at Wabamun and at a natural nest site in northern Alberta (Dekker 1999), may seem to refute the above hypothesis, because there can be no doubt that adult females are accomplished hunters. Therefore, the male's reason for making live-drops to his mate can have nothing to do with teaching. However, here a secondary and complementary factor comes into play. Whether hunting with his mate or a fledgling, the male plays a dual role in support of both.

It is difficult if not impossible to translate all of the anecdotes detailed in the appendix into hard data in support of the teaching hypothesis. As to the second hypothesis that parent Peregrines assist their fledglings in capturing their first prey, on the evidence presented here, we are convinced that the answer is affirmative. In the final analysis, there is probably more going on in the mind and understanding of Peregrines than a researcher can prove.

APPENDIX

Observations of parental care by a pair of Peregrines toward their fledglings and nearly fledged juveniles.
(**1**) On 16 August 1999, having stooped at a gull and missed his target, the adult male manoeuvred under it in a peculiarly slow flying style. He appeared to drive the gull upward and prevent it from taking refuge in the water below. Seconds later, a juvenile female stooped from above, just missing the gull. The adult male then stooped again, just short of the target. Chased by both Peregrines, the gull eventually escaped. In this case, as in others, it seemed clear that the male did not want to catch the gull himself. Instead, he attempted to set it up for the approaching juvenile.
(**2**) On two occasions, when a gull was chased by a combination of one adult and one fledgling, the adult smacked or grazed the prey, causing it to cartwheel in

the air. In the next instant, the gull was seized by the closely following juvenile. Our strong impression was that the adults were not intent on taking the gull themselves.

(**3**) On 12 August 2000, the adult male, while soaring in tandem with his mate, suddenly sprinted away (400–500 m) to hit a shorebird that was being chased by two juveniles at an altitude estimated at 75 m. Dead or stunned, the prey fell into reeds and was lost before the juveniles could recover it. This was one of only four instances in which these Peregrines smacked their prey in the air, as opposed to the common method of binding to the prey and carrying it down. The remarkable point is that the adult male did not stoop and recover the plunging shorebird himself, which would have been a normal procedure if he had been hunting alone.

(**4**) A Rock Dove pursued by a fledgling was taken by the adult female, who transferred it at once to the juvenile. On their own, the adults seldom pursued doves.

(**5**) Fledglings that have been on the wing for only a few days have difficulty carrying a gull. After making an aerial transfer to an unskilled juvenile, the adults will accompany it in flight and retrieve the gull if dropped. By the same token, seemingly aware of the lack of ability in a badgering female youngster, the adult female refused to transfer a just-caught gull and instead carried it circa 500 m to the roof of the plant, where she at once surrendered it to the fledgling.

(**6**) The adult male demonstrated an awareness of the different needs of the sexes. In group hunts with male offspring he selected mostly passerines or small shorebirds (41 of 45). By contrast, in the company of females he usually attacked gulls. After capturing one or more gulls for his mate or juveniles, he hunted smaller prey for himself.

The above interactions between parents and fledglings parallel numerous anecdotes detailed by long-term Peregrine observers in New York and Great Britain (Frank 1994; Treleaven 1998). It is our strong impression that adult Peregrines appear to know how to assist their progeny in gaining independence. For a start, there is a change in food exchanges when the first of the juveniles are close to fledging. Both at Wabamun and a natural nest site in central Alberta, the adult male appeared to coax a fully feathered male youngster to fly. Instead of transferring the prey item in the usual direct manner, he cruised back and forth just out of reach, holding the prey in his lowered feet. His behaviour cannot be explained as having been prompted by a fear-based reluctance to approach an aggressive female for this incident involved a male juvenile. After the young male failed to fledge, the adult gave the food item to another juvenile, a fully feathered female sitting on the catwalk. Similarly, on two other occasions when he arrived with a small prey, he flew right up to a screaming juvenile male that was perched on the edge of a catwalk, but instead of exchanging the prey feet-to-feet, he dropped it and recovered the falling item in a quick stoop. He repeated this teasing show three or four times. After the fully feathered juvenile male failed to fly, the adult gave the food item directly to a less advanced female still in the nest box.

Another noteworthy exchange involved the female. One day, after she had delivered a prey to two juvenile males, which shared the food, the male arrived, also with prey. He gave it directly to one of the same two youngsters. Instantly, the adult female interfered. She took the food item away from the young male and offered it instead to a juvenile female, which had not been fed for several hours. Similar incidents were seen at Wabamun and at a natural nest site in central Alberta (D. Dekker, unpublished data).

Table 10.1. Hunts and kills made by a pair of Peregrine Falcons nesting on a powerplant in central Alberta, 1998–2004. The hunting success rates are in parentheses.

	Hunts / kills from soaring		Hunts / kills from perch		Total hunts	
Adult male	176/44	(25.0%)	40/8	(20.0%)	216/52	(24.1%)
Adult female	52/18	(34.6%)	18/8	(44.4%)	70/26	(37.1%)
In Tandem	68/23	(33.8%)	32/16	(50.0%)	100/39	(39.0%)
Totals	296/85	(28.7%)	90/32	(35.5%)	386/117	(30.3%)

Table 10.2. Prey taxa captured by the Peregrine pair hunting solo or in tandem.

	Franklin's Gulls	Small Passerines	Small Shorebirds	Ducks	Other or Unidentified
Adult male	14	27	1	1	9
Adult female	21	2	1	2	–
Pair in tandem	27	3	5	–	4
Totals	62	32	7	3	13

Table 10.3. Hunts/kills made by one or both parents accompanied by one or more fledgling juveniles.

	1 Juvenile	2 Juveniles	3 Juveniles	Totals
Adult male	29/10	34/10	6/2	69/22
Adult female	3/2	6/4	–	9/6
Both adults	6/3	5/0	1/0	12/3
Totals	38/15	45/14	7/2	90/31

Table 10.4. Prey captured as a percentage of total prey per year.

Year	% Gulls	% Passerines	% Other	# Total Prey	# Hunts
1	8.7	60.9	30.4	23	108
2	63.6	27.3	9.1	11	47
3	62.5	12.5	25.0	8	25
4	58.3	16.7	25.0	12	37
5	43.7	31.3	25.0	16	44
6	63.6	27.3	9.1	22	61
7	80.0	4.0	16.0	25	64
Totals	53.0	27.4	19.6	117	386

Table 10.5. Comparison of main prey taxa captured in years #1 and #2-7.

	Prey taxa	# Kills	% of total	n
Year #1	Gulls	2	8.7	23
Years #2-7		60	63.8	94
Year #1	Passerines	14	60.9	23
Years #2-7		18	19.1	94

Table 10.6. Number of hunts by the adult male and female, either hunting solo or in tandem, in year #1 and years #2-7.

		# Hunts	% of total
Year #1	Adult male	85	78.7
$n = 108$	Adult female	3	2.7
	Tandem	18.5	
Years #2-7	Adult male	131	47.1
$n = 278$	Adult female	67	24.1
	Tandem	28.8	

Figure 10.1. Hunting success rates per year of the same pair of Peregrine Falcons breeding in central Alberta, 1998--2004. The respective number of hunts and kills are: 108/23; 47/11; 25/8; 37/12; 44/16; 61/22; 64/25.

(Footnote: In 2005–2007, continued observations resulted in an additional 54 hunts and 25 kills, which represents a further increase in hunting success rate during this 3--year period to 46.3%. The female falcon was the same individual in all years. In 2007, she was 12 years old. Another change we noted was that the adult male began to feed on gulls as well. This change in his prey preference might have been linked to increasing scarcity of small passerines in the area.)

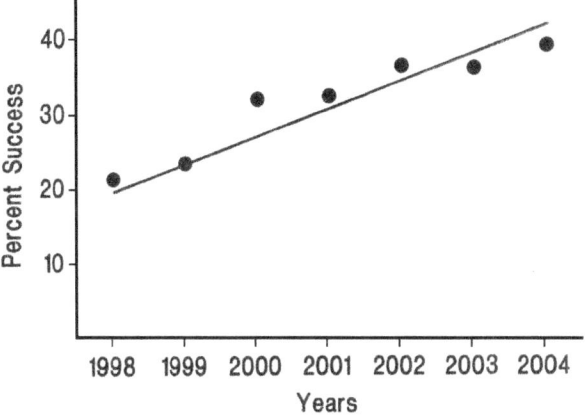

CHAPTER 11

A comparison of Peregrines and Merlins hunting small shorebirds and passerines.

Dekker, D. 1988. Peregrine Falcon and Merlin predation on small shorebirds and passerines in Alberta. Canadian Journal of Zoology 66:925-928.

Abstract – The hunting techniques and predatory efficiency of the Peregrine Falcon (*Falco peregrinus*) and the Merlin (*Falco columbarius*) preying on small shorebirds and passerines were divided into four categories: (1) surprise attacks with less than six swoops at prey flushed from the ground or water; (2) pursuits with less than six swoops at flying prey; (3) persistent, long-range pursuits with more than six swoops at flying prey; and (4) other or unknown methods. The surprise method made up, respectively, 73.9% of 647 hunts by Peregrines, and 72.3% of 354 hunts by Merlins. Persistent pursuits of prey accounted only for 2.5% of Peregrine hunts and 4.5% of Merlin hunts, but they were significantly more successful than surprise attacks for the Merlins (62.5% vs. 9.4%, $P < 0.00001$). Persistent pursuits were also more successful than surprise attacks for Peregrines, but the difference was not significant (18.8% vs. 8.8%, $P = 0.40$). In this paper, I postulate that Peregrines and Merlins employ persistent hunting techniques only if surprise attacks are unproductive because of (a) lack of expertise of the falcons, (b) scarcity of prey, (c) unsuitable habitat, and (d) effective early warning by flocking prey. Peregrines captured 50 shorebirds of 11 species, and three passerines of two species. Merlins captured 28 shorebirds of eight species, and 16 passerines of five species. Adult male Merlins were significantly more successful than immatures and adult females combined (38.7 vs. 9.1%, $P < 0.00001$). Adult Peregrines were not significantly more successful than either spring or fall immatures (9.2% vs. 7.2% and 4.9%, $P = 0.21$, $P = 0.24$, respectively). Merlins were significantly more successful than Peregrines in hunting small passerines (12.2 vs. 3.8%, $P < 0.05$), but not significantly more successful in hunting small shorebirds (12.5% vs. 8.8%, $P = 0.10$).

INTRODUCTION

Although the weight and wing loading of the smallest Peregrine Falcons (*Falco peregrinus*) are more than twice as great as those of the largest Merlins (*Falco columbarius*), the range of prey taken by the two species overlaps widely (Brown and Amadon 1968). Peregrine predation on small passerines and shorebirds has been documented from North America, Europe, and Africa (Dekker 1980; Ratcliffe 1980; Thiollay 1982). Merlin predation on the same group of birds has been reported from Europe and California (Rudebeck 1950; Page and Whitacre 1975). However, the predatory habits and efficiency of the two falcons should not be compared unless the locality, time of year, and prey species are the same. This study describes and compares the foraging habits and hunting success of Peregrines and Merlins preying on small migrant shorebirds and passerines in central Alberta during spring and fall.

STUDY AREA AND METHODS

The study area was a strip of ca. 15 x 2 km of pastures and fields adjacent to Beaverhills Lake near Edmonton, Alberta ($53°$ N, $112°$ W). Shorebirds and open-country passerines were numerous during spring and fall. Migrating Peregrine Falcons occurred from mid-April to late May, and from early September to early October (Dekker 1984). Merlins, either migrating through or locally breeding, were seen from mid-April to late November.

From 1965 to 1987, the study area was visited on about 24 days each spring (15 April to 31 May) and on about 12 days each fall (1 September to 15 October), for totals of 549 and 265 days. Falcons were spotted by frequently scanning through binoculars or telescope. Flying falcons were observed for as long as they remained visible. Resting falcons were often watched until they flew away of their own accord. Observations were recorded in notebooks.

A hunt is defined as a completed attempt by a falcon to seize prey including one or more swoops at the same target. Hunts that resulted in capture were termed successful even though a few preys escaped or were released alive after capture. The term stoop refers to a falcon's sudden descent at high speed with wings flexed and rigid. The term swoop is used for the abrupt downward shift in direction with beating wings that falcons perform when closing with and attempting to seize flying prey. Hunting methods were divided into four categories: (1) surprise attacks with or without a short pursuit and less than six swoops at prey initially on the ground or in shallow water; (2) short-range pursuits with less than six swoops at flying prey; (3) persistent long-distance pursuits with more than six swoops; and (4) other or unknown methods.

Adult Peregrines were distinguished from immatures by dorsal colouration and ventral markings. Adult female Merlins could not be reliably distinguished from immatures, but adult male Merlins were identified from all others by their blue-grey dorsal colour.

Data sets were compared statistically by x^2 contingency tests.

RESULTS

Peregrine Falcons attacked prey species in a total of 1478 hunts, but the outcome of 412 was unknown. Of the remaining 1066 hunts, 569 were directed at shorebirds, resulting in 50 kills of 11 species, and 78 were aimed at small passerines, resulting in three kills of two species; 419 hunts involved other or unknown prey. Merlin hunts with known outcome totalled 360, of which 223 involved shorebirds, resulting in 28 kills of eight species, and 131 hunts were directed at passerines, resulting in 16 kills of five species. Six Merlin hunts involved unknown prey species (Tables 11.1 and 11.2).

The majority of Peregrine and Merlin hunts (73.9% and 72.3%, respectively) were surprise attacks on prey that was on the ground or in water. Peregrines began these attacks from distances of over 0.5 km and from altitudes varying from less than 1 m to >1 km. In 220 cases, they launched surprise attacks from high soaring flight. For detailed descriptions of Peregrine hunting methods, see Dekker (1980, 1987). Merlins generally began surprise attacks from much shorter distances and usually started from a low perch or sustained flapping flight low over the ground, rarely from high flapping or soaring flight. Surprise attacks by high soaring Peregrines were initiated with a stoop that could be either vertical or oblique.

Surprise attacks from sustained flapping flight by Merlins as well as Peregrines were accompanied by increased wing beats. In the last stage of these hunts, both species of falcon flew very low above the ground or water and attempted to seize the prey just after it flushed. Surprise could be complete, especially if the falcon's approach was screened by vegetation. Of 66 prey caught in surprise attacks by either Peregrine or Merlin, 40 (60.6%) were seized at once. If the intended target flushed in the nick of time, it often dodged the falcons by instantly dropping back onto the ground or into the water, causing the falcon to overshoot its target. In the next instant, most birds took off again away from the falcon, but some did not rise if the falcon was overhead. The falcon – either Peregrine or Merlin – then stooped or swooped down, attempting to seize the prey on the ground or lift it from the water. I saw no evidence that shorebirds or passerines were hit in the air and knocked down, although the plunging escape tactics of some prey made it look as if they had been struck down. Aerial "knock-downs" were also the exception for ducks captured by Peregrines in the study area (Dekker 1980, 1987).

(However, after the termination of this study, during the spring migration period and at very close range, I saw an adult female Peregrine stoop at a flushing flock of Semipalmated Sandpipers (*Calidris pusilla*). One of them dropped out, splashing into the water. The Peregrine retrieved the evidently dead bird from the surface, carried it to a shoreline stone, but left the carcass uneaten. Aerial hits of prey were somewhat less rare for Peregrines hunting Dunlins (*Calidris alpina*) wintering on the Pacific coast of British Columbia.)

Shorebirds that successfully dodged the initial first pass of a surprise attack took off at high speed and evaded the pursuing falcon by making sharp turns and

Adult male Merlin of the pale central Canadian subspecies (F. c. richardsonii). The brown plumage of adult females resembles that of all first-year Merlins. (Photo Gerald Romanchuk).

twists. Both Peregrine and Merlin were capable of overtaking and outclimbing sandpipers, but the Merlin was evidently much better at following the erratic prey, losing less altitude and distance between swoops. By contrast, the Peregrine's swoop gave the impression of being "over-powered" and it often widely missed the target. However, as long as the falcon could regain a height advantage, it quickly swooped downward again to overtake its prey. After dodging repeated swoops, sandpipers and passerines were eventually caught or let go. As a last resort, they descended steeply, followed by the stooping falcon, to find cover in bushes or reeds.

Persistent, long-distance pursuits with >6 swoops accounted for 2.5 and 4.5% of all Peregrine and Merlin hunts, respectively. Persistent hunts might cover distances of >2 km, and the outcome of 20–25 chases could not be ascertained, voiding these data for inclusion in the tabulations of this study. In nine persistent hunts, two or more Peregrines pursued and swooped alternatively at sandpipers until they were either caught or escaped. Tandem hunts with alternating swoops were common for Peregrines, but very rare for Merlins. I saw only two pairs of non-breeding Merlins chasing sandpipers over long distances, of which one ended in a capture. The success rate of persistent hunts as compared to surprise hunts was significantly higher for Merlins ($P < 0.00001$) but not for Peregrines ($P = 0.40$).

The success rate of Peregrines that could be recognized as either adult or immature, hunting all prey species, was 9.2% ($n = 305$) for adults, 7.2% ($n = 549$) for spring immatures, and 4.9% ($n = 61$) for fall immatures, but the differences were not significant ($P = 0.21$ and $P = 0.24$, respectively). Adult male Merlins had a 38.7% ($n = 31$) success rate, significantly higher than the 9.1% ($n = 331$) of brown-backed Merlins (immatures and adult females).

Peregrines consumed and carried away all but one prey they captured. Merlins released one Lesser Yellowlegs (*Tringa flavipes*) alive and abandoned one just-killed Semipalmated Sandpiper (*Calidris pusilla*). Peregrines that had captured prey were often chased and sometimes robbed by conspecifics, and Peregrines robbed Merlins (Dekker 1980). Merlins carrying prey were chased by Swainson's Hawks (*Buteo swainsoni*) and Red-tailed Hawks (*Buteo jama-icensis*).

DISCUSSION

The majority of hunts (73.3%) were surprise attacks, which should be an energy-efficient method of foraging, yet 91.0% were unsuccessful. Ratcliffe (1980) questioned the intent of unsuccessful hunts by Peregrines and suggested that the falcons may have been playing or practicing. Dekker (1980) hypothesized that the success rate of hunting Peregrines depends on expertise as well as motivation, and that adult Peregrines feeding young should have a higher success rate than migrating or wintering falcons. Roalkvam (1985) confirmed that hypothesis in a review of existing literature. He calculated that the hunting success rate of breeding Peregrines was significantly higher than that of non-breeding Peregrines, and higher in adults than in immatures. In this study, migrating adult

Peregrines were not significantly more successful than first-year Peregrines, but adult male Merlins, which were on breeding territory in the study area, were significantly more successful than the sample of hunts by all brown-backed Merlins (immatures and adult females combined).

The need to forage for nest young and/or a mate probably motivates a breeding falcon to hunt with greater intensity than do non-breeding falcons. Other likely motivating factors are hunger and the natural impulse to exploit an unusual easy opportunity that happens to present itself. I postulate that highly motivated falcons use the most productive hunting methods available to them, such as persistent pursuit, if they are unable to succeed in short-range pursuits, or if they are prevented from making successful surprise attacks. I further postulate that the factors preventing productive surprise attacks are: (a) lack of expertise of the falcon; (b) scarcity of prey; (c) unsuitable habitat; and (d) effective early warning behaviour of prey. These four conditions are illustrated by the following observations.

In this study, all persistent pursuits by Peregrines involved first-year falcons. I suggest that these immatures were forced to hunt with persistence because they lacked the skill to succeed in surprise attacks. They appeared to have sufficient speed to get close to prey in surprise hunts, but they usually failed at the critical moment when the quarry flushed and dodged the attack. Once the immature had lost its "first strike" advantage, the prey got away. Hunt et al. (1975) observed long pursuits of prey by first-fall Peregrines migrating along the Texas coast. In Alberta, I often saw long chases and clumsy attempts to seize prey that had ditched into water, some involving >30 swoops, while the target birds evaded each pass by splashing and diving repeatedly. Some prey managed to rise and get away again. By contrast, adult falcons doubled back in an instant and seized such dodging prey at the first attempt. One adult Peregrine briefly hovered over a submerged dowitcher (*Limnodromus* spp.) that had dived into floodwater and retrieved it from the bottom (Dekker 1980). In the process, the falcon went belly deep into the pond.

A falcon with only scarce opportunities to capture prey will probably be motivated to hunt with more persistence than a well-fed falcon if prey is abundant in the environment. For instance, a most persistent hunt by an adult male Merlin occurred on a day when spring migration was over and I had seen only a single sandpiper flying by over the lake. An adult Merlin which I had been watching for some time left its tree perch and pursued the sandpiper vigorously, catching it on the 28th swoop, the longest series I have thus far recorded for adult falcons of either species. During summer, small prey species are scarce in the study area and resident passerines stay close to cover. It was common to see Merlins chase locally breeding sparrows (Emberizinae) until they plunged into long grass. In a few cases, when I flushed these hiding sparrows, a Merlin immediately pursued them again. Blackbirds (Icterinae) commonly escaped from pursuing falcons by taking cover in reeds or under grazing cattle. In semi-open parkland away from the study area, I saw breeding Merlins make long ascending flights, launched from trees, to intercept high-flying passerines, all of which plunged into woods just before being overtaken. Hunt et al. (1975) pointed out the advantages of open beach habitat for foraging Peregrines during migration. Bird and Aubrey (1982) and Thiollay (1982) reported high success rates (25–35%) for breeding Peregrines hunting land birds over water well away from tree cover.

Flocking of prey species is considered an anti-predator strategy (Powell 1974). Lone birds or isolated small flocks are more vulnerable to Peregrine predation than prey individuals in large flocks (Dekker 1980; Thiollay 1982). On days when thousands of sandpipers had gathered on an open pasture to feed on a massive hatch of midges (Chironomids), I saw adult Peregrines repeatedly fail in their attempts to catch the sandpipers by surprise. Most of their attacks were aborted after the intended quarry flushed well ahead, evidently warned by alarm signals of flocks that were not under direct attack but had spotted the falcon pass by. The Peregrines eventually left the area, unable to make a capture despite – and probably because of – the local multitude of birds.

To investigate the hunting habits of birds of prey, it is important that the researcher focuses on more than a single subject of the species under study, since it is possible that individual birds have evolved particular habits that are not shared by their conspecifics. In this study, Merlins were successful in 17 (23.3%) of 73 aerial pursuits. By contrast, Page and Whitacre (1975), who studied a single adult female Merlin wintering in California, reported that this bird failed in all of 82 pursuits of flying prey. Its overall success rate in 344 hunts of small shorebirds and passerines was 12.8%, and surprise methods made up 80% of all hunts, which is nearly identical with the results reported here. The hunting success rates of Peregrines in this study fall within the range of 7.3% – 13.7% reported for migrating and wintering Peregrines in three Swedish areas (Roalkvam 1985).

In summing up, the hunting methods used by Peregrines and Merlins preying on small shorebirds and passerines are very similar and depend largely on a stealth approach to take the prey by surprise. Because of its superior manoeuvrability and lighter wing loading, the Merlin was significantly more successful than the Peregrine as a predator of passerines (12.2% vs. 3.8%, $P < 0.05$), but not significantly more successful as a hunter of small shorebirds (12.6% vs. 8.8%, $P = 0.10$).

Table 11.1. Strategies, hunts/kills, and success rates of Peregrine Falcons (P) and Merlins (M) preying on small shorebirds and passerines in Alberta.

	Shorebirds		Passerines		Totals (% success)			
	P	M	P	M	P		M	
Surprise attacks	439/40	192/20	39/2	64/4	478/42	(8.8)	256/24	(9.4)
Short pursuits	81/3	13/1	30/1	44/6	111/4	(3.6)	57/7	(12.3)
Persistent pursuits	13/3	8.6	3/0	8/4	16/3	(18.8)	16/10	(62.5)
Other methods	36/4	10/1	6/0	15/2	42/4	(9.5)	25/3	(12.0)
Totals		569/50 223/28	78/3	131/16	647/53	(8.2)	354/44	(12.4)

Table 11.2. Species and numbers of shorebirds and passerines seen to be captured by Peregrine Falcons and Merlins in Alberta.

Prey species captured by	Peregrines	Merlins
Black-bellied Plover (*Pluvialis squatarola*)	2	–
Lesser Golden Plover (*Pluvialis dominica*)	1	–
Lesser Yellowlegs (*Tringa flavipes*)	5	1
Sanderling (*Calidris alba*)	1	2
Semipalmated Sandpiper (*Calidris pusilla*)	2	2
Least Sandpiper (*Calidris minutilla*)	1	2
Baird's Sandpiper (*Calidris bairdii*)	3	1
Pectoral Sandpiper (*Calidris melanotos*)	21	6
Stilt Sandpiper (*Calidris himantopus*)	*	2
Buff-breasted Sandpiper (*Tryngites subruficollis*)	1	*
Dowitcher (*Limnodromus* spp.)	2	–
Wilson's Phalarope (*Phalaropus tricolor*)	1	–
Red-necked Phalarope (*Phalaropus lobatus*)	*	1
Unidentified sandpipers	6	11
Unidentified shorebirds	4	–
Horned Lark (*Eremophila alpestris*)	–	1
Savannah Sparrow (*Passerculus sandwichensis*)	1	*
Snow Bunting (*Plectrophenax nivalis*)	2	6
Brown-headed Cowbird (*Molothrus ater*)	–	1
Unidentified swallow	–	1
Common Redpoll (*Carduelis flammea*)	–	2
Unidentified passerines	–	5
Totals	53	44

*Found feeding on this species but capture not observed

Peregrines also captured 27 ducks of nine species, one Franklin's Gull (*Larus pipixcan*), and one Common Tern (*Sterna hirundo*) (Dekker 1980, 1987).
Merlins also captured one flightless Lesser Yellowlegs, two juvenile passerines, numerous dragonflies (*Odonata*) and one bat (*Myotis lucifugus*).

CHAPTER 12

Nest site competition between Peregrines and Prairie Falcons

Dekker, D. and R. Corrigan. 2006. Population fluctuations and agonistic interaction of Peregrine and Prairie Falcons in central Alberta, 1960–2006. The Journal of Raptor Research 40:255–263.

Abstract – The Peregrine Falcon *(Falco peregrinus)* became locally extirpated during the 1960s while the Prairie Falcon (*Falco mexicanus*) increased in a sympatric nesting range along a section of river in Alberta, Canada. Following the release of 233 captive-reared juvenile Peregrines in 1992–1996, some Peregrines returned to the river to breed, reaching a high of seven pairs in 2000, but declining again to two breeding pairs in 2006. In an inverse relationship, the number of Prairie Falcon nests dropped from five to two and then went back up to four. We hypothesize that the Peregrine's failure to reach former densities in the study area was due to habitat deterioration, which resulted in a reduced prey base compared to pre-1960s conditions. Interspecific strife with Prairie Falcons was common at contested nest sites. The Peregrines were the dominant aggressors, replacing pairs of Prairie Falcons at some, but not all, historical cliff eyries.

INTRODUCTION

The North American breeding range of the Peregrine Falcon (*Falco peregrinus*) has the entire geographic range of the Prairie Falcon (*Falco mexicanus*) within it (Palmer 1988). The two species are quite similar in size and physiology. However, they are ecologically separated by habitat and prey preference, with the Peregrine mainly selecting sites near water (Cade 1982), and the Prairie Falcon inhabiting arid regions (Enderson 1964; Brown and Amadon 1968; Steenhof 1998). Where they occur sympatrically, the Prairie Falcon is usually the more common of the two (Porter and White 1973).

Food availability is an important criterion in the distribution and abundance of the Peregrine in its nearly worldwide range (Ratcliffe 1980). In the American West, scarcity of prey and competition for prey with the Prairie Falcon may contribute to the relative paucity of the Peregrine in arid regions (Porter and White 1973). In addition, they compete for cliff nesting sites, and where suitable locations are scarce, the presence of one species may influence the numbers and distribution of the other (Newton 1979). Nevertheless, the Peregrine and the Prairie Falcon are reported to be more or less compatible and coexist side by side with a minimum of hostilities, even sharing the same set of cliffs and switching nesting sites in different years (Porter and White 1973; Cade 1982; Palmer 1988).

The published literature contains very little and only anecdotal information on agonistic interactions between Peregrines and Prairie Falcons. In aerial performance, the latter is considered a match for the Peregrine (Walton 1978; Cade 1982), and at higher elevations, the Prairie Falcon – apparently due to its lighter wing loading – is believed to outfly the Peregrine (Brown and Amadon 1968: 835; Nelson 1969). However, in a few cases, Peregrines are known to have replaced Prairie Falcons at nest sites and forced them to abandon their eggs (Porter and White 1973:60). In California, a female Prairie Falcon was struck down and assumed to have been killed in a cooperative attack by a pair of Peregrines (Walton 1978). In Utah, Peregrines killed the three young of Prairie Falcons soon after fledging from their nest 300 m away from the Peregrines (White et al. 2002). By contrast, there are no reports of Peregrine fatalities caused by Prairie Falcons. Interspecific kleptoparasitism (the robbing of prey by one species from the other) may be expected to occur either way, but in the only two published incidents known to us the Peregrine was the loser (Beebe and Webster 1964; Dekker 1999).

Changes in climatic conditions or land management that result in a loss of essential habitat and food resource may favour one falcon species over the other. For example, during a period of drought on a sympatric breeding range in Utah, about 50% of the Peregrine eyries became vacant and were taken over by Prairie Falcons (Nelson 1969). A similar switch occurred in central Alberta during the 1960s when the Peregrine population died out (Enderson 1969). The primary cause of this extirpation was believed to be toxic chemical residues in the food chain, superimposed on the robbing of all local nest young by humans (Dekker 1967). Less clear was the detrimental impact of environmental changes, such as modern agricultural practices and the draining of area wetlands. During the process of Peregrine extirpation in central Alberta, the number of Prairie Falcons increased, suggesting the possibility that the latter had displaced the Peregrines by aggressive competition for prey or nest sites (Dekker 1967). An effort to better understand the relative importance of the above factors was recommended at the conclusion of the 2000 Canadian Peregrine Falcon Survey (Rowell et al. 2003).

An opportunity to re-examine the interspecific dynamics arrived between 1992–1996 with the reintroduction of captive-reared Peregrines into the study area by the Alberta Fish and Wildlife Division (AFWD) in association with the Canadian Wildlife Service (CWS). Similar programs have been very successful across North America (Cade and Burnham 2003). In this paper, we report on: (1) the number of nest sites occupied by Peregrine Falcons in the study area before their 1960s extirpation; (2) Peregrine population fluctuations after their

1992–1996 reintroductions; and (3) agonistic interactions between Peregrines and Prairie Falcons at contested nest sites.

STUDY AREA AND METHODS

The study area consists of two sections (A and B), respectively 30 and 100 km long, of the Red Deer River valley in central Alberta, Canada (52°N, 112°W). The river is 80–150 m wide and flows in a southeasterly direction through a gently sloping agricultural plain with widely scattered farm buildings and woodlots. Depressions are filled with shallow wetlands that vary in size depending on annual precipitation. The valley is 0.5–1 km wide and about 80 m lower than the mean elevation (860 m) of the uplands. Portions of the valley bottom are used for agriculture and cattle grazing. South-facing valley walls are semi-arid and grassy; the north-facing aspect supports a narrow strip of white spruce (*Picea glauca*) and trembling aspen (*Populus tremoloidus*). Steep and eroding escarpments feature limestone outcrops and cliffs. The local climate is cold continental, and the river is ice-covered from November to late March.

Dependent upon suitable habitat, the original and current nesting range of the Peregrine includes all of Alberta (Palmer 1988; Corrigan 2000). The Prairie Falcon's breeding range in North America reaches its northern limit at the latitude of the study area (Steenhof 1998), although there are two isolated nesting records, respectively 150 km (Semenchuk 1982) and 500 km farther north in Alberta (R. Corrigan, AFWD, unpublished data). While the Peregrine Falcon migrates south in early fall, the Prairie Falcon is a year-round resident in the province (Godfrey 1986; Dekker and Lange 2001). Breeding pairs of Peregrines arrive in the study area in the first two weeks of April. By that time, the local Prairie Falcons are well established and their reproductive season is roughly three weeks ahead of that of the Peregrines (Dekker, unpublished data).

Study area A was first surveyed by DD, on foot and by canoe, in 1960. Repeat visits (5–7/year) were made to a 4-km central section, named Valley of the Falcons (VF), which contained five prominent nesting cliffs. Area A was not visited in 1962–1964. Seasonal checks resumed in 1965–1969, but dropped to occasional in 1969–1996. The frequency of day visits to the VF section increased again in 1997–2006 to a total of 72 days (3–11/year). The nest sites were in plain view from the opposite side of the river and could be watched through binoculars and telescope. Observation time spent at one or more falcon eyries in VF was 3–10 hours/day with a mean of 7 hours/day.

Study area B began downstream from area A and was first visited in 1992, when AFWD began a mass release of captive-reared Peregrine Falcons. During the program's five years of operation, hack boxes were installed on four area B cliffs, two of which were occupied at the time by Prairie Falcons. Their young were removed and taken into captivity. In 1996, a fifth release box was installed in area A at a site which had no known occupancy – past or present – by wild falcons of either species. The total number of juvenile Peregrines successfully released on the Red Deer and other southern Alberta Rivers was 223 (Holroyd

2003). After the releases had stopped, RC travelled the river by jet boat and helicopter to locate nesting Peregrines. Known sites were monitored yearly. One or more captive-reared young were added to four pairs with less than four young of their own. At about three weeks of age, nest young were fitted with standard metal and plastic alpha-numeric bands: red for captive-reared birds and black for wild-raised young.

Opportunities for studying the foraging habits of these falcons were limited. We rarely saw them attack prey. Upon leaving the valley, both the Peregrines and the Prairie Falcons commonly soared and sailed out of view (Dekker 1993, 1999). Waiting for their return, we recorded the number and species of prey items brought back to the nest. We did not collect prey remains at the eyries.

RESULTS AND DISCUSSION

In 1960, study area A contained five nesting pairs of Peregrines, including two in the VF section. Single adults were seen at two additional sites up river from VF. In 1965, two of the known nesting sites were still occupied; one situated in the VF section, the other 10 km downriver from there. Both had been abandoned by 1967 (Dekker 1993).

Area A contained four breeding pairs of Prairie Falcons in 1960, and three in 1965. In addition, two single Prairie Falcons frequented VF cliffs after they had been vacated by Peregrines.

We have no information on the historical Peregrine and Prairie Falcon population of study area B. No Peregrines were found in 1992 when AFWD and CWS personnel began surveying for cliff sites suitable for the release of captive-reared Peregrines. The number of Prairie Falcons was likely greater than the two pairs recorded at cliffs selected for the placement of nest boxes. In 1996, after the Peregrine releases were completed, first-year immatures, returning from their migrations, became summer residents at the hack sites. In 1994 the first mated pair laid eggs and raised young in one of the boxes. Henceforth, up until 2006, one or more of the hack locations were occupied yearly, although the falcons sometimes laid their eggs on nearby ledges instead of inside the boxes.

In 1997, RC found the first successful nest site at a natural cliff in the valley. This was the very last cliff that had been used by Peregrines prior to their extirpation (Dekker 1999). By 2000, there were pairs at three traditional natural sites in area A, and the total number of Peregrine nests in area A and B reached a high of seven (Figure 13.1).

The Peregrine expansion ended in 2000 and the population began to decline. By 2006, the number of eyries had dropped from seven to two (Figure 12.1). One of these pairs used an area B nest box, while area A contained the only productive nest on a natural cliff ledge. The mean number of fledglings (excluding fostered young) was 1.8 per occupied site ($n = 54$), which is near the average for many Peregrine populations (White et al. 2002). However, the turnover of one or both adults was high at eyries that were occupied for two or more years in a row (8/12), and there was a rather high incidence of first-year females ($n = 4$) in nesting pairs which failed to produce young. The presence of first-year recruits on nest sites is typical of populations that are in flux (Newton 2003; Tordoff and Redig 2003), and Peregrines are known to do poorly in their first breeding attempts (Court et al. 1988b).

The Prairie Falcon population in area A showed an inverse relationship with the Peregrine fluctuation (Figure 12.2). After 1995, the number of resident Prairie Falcon pairs declined from five to two during the years of the Peregrine expansion, but increased again to four in 2004–2006. The mean number of young at successful Prairie Falcon eyries was 4.4 ($n = 25$), but not all sites were checked after egg-laying time. In most years, two Prairie Falcon nestlings were collected under falconry permits. In three instances, entire broods disappeared for unknown reasons.

After 1995, only one VF site has been continuously in use by Prairie Falcons nesting in the same cavity each year. The four other sites alternated in occupancy between Prairie Falcons and Peregrines that used the same or different nest ledges. Aggressive interactions between neighbouring pairs were observed only at the two VF sites that were closest together (400 m), but only when the pairs were of different species, and only during the early part of the breeding season. Area A cliffs known to have been occupied over the years by breeding falcons of either species varied in aspect; five were north-facing and four faced south. By 1995, four of five occupied sites faced north.

Interspecific strife. We recorded agonistic interactions between Peregrines and Prairie Falcons at five different cliff sites on 33 days. The Peregrines always were the aggressors and appeared to be faster, more agile and more persistent in pursuit than the Prairie Falcons. In contrast, the latter allowed their opponents to pass by unchallenged. Frequent aerial combat took place at only three cliffs. In years when both species were present early in the breeding season, the Peregrines dominated the air space above, below and on either side of the Prairie Falcons' nest ledges. Prey-carrying parent Prairie Falcons that were intercepted by Peregrines evaded their attackers by dropping down onto the escarpment, often into some vegetation. Bodily contact was seen only once, when a male Prairie Falcon was seized by a male Peregrine and forced down to the ground. Nevertheless, the pair of Prairie Falcons prevailed and remained in possession of the superior nest ledge.

We recorded four conflicts in which a pair of Peregrines succeeded in evicting Prairie Falcons from their nest sites. However, in four other cases the Peregrines were unable to evict the Prairie Falcons, and in two additional cases a site usurped by Peregrines reverted back to Prairie Falcons the subsequent year. The appendix contains details of ownership changes at three of the most contested sites.

As falcons tend to "use the same nesting places in different years, monitoring becomes easier and the value of population studies increases the longer they are continued" (Newton 1979:55). Although our sample size is small, the long-range data for study area A show that the reintroduced Peregrines regained 60% (three out of five) of their historical nesting sites. Moreover, they succeeded in their comeback in a relatively short time of three years. However, contrary to expectations (Holroyd and Banasch 1990), the successful reintroduction of the Peregrine did not lead to continued expansion along the Red Deer River. We can only speculate as to the cause or causes.

Nest site competition. On the basis of our observations that the Peregrine was the dominant aggressor with superior flying skills at all contested nest sites, we reject the hypothesis that the 2000–2006 Peregrine declines were the result of interspecific competition with the Prairie Falcon for the nest sites. However, such competition may have been a contributing factor, as suggested by the fact that the Peregrines often failed to form pairs and that they were not always successful in driving off nesting pairs of Prairie Falcons.

We have no proof that the returning Peregrines were subjected to human-caused mortalities on the study area, nor do we have any reason to suspect that agricultural or industrial chemical residues were implicated in the current population decline. As noted, the incidence of first-year recruits on nest sites was rather high, which is typical of populations that are in flux (Newton 2003; Tordoff and Redig 2003), and Peregrines are known to do poorly in their first breeding attempts (Court et al. 1988b).

However, by comparison, most other Peregrine reintroduction programs have resulted in sustained growth (Enderson 2003). Even in regions where no releases took place, contemporary Peregrine populations have regained or are now exceeding historical densities (Cade and Burnham 2003). Philopatry and site tenacity are typical in Peregrine ecology, and a numerical increase in recovering populations is an indicator of population viability and normalcy (Cade 2003). In contrast, the absence of such growth in central Alberta is notable and it spurred us on to investigate what might be happening there.

The prey base hypothesis. The introduced Peregrine population in the study area began its downward adjustment four years after the releases were stopped. Apparently, while the captive-reared cohort was dying off, it was not being replaced at the same rate by a younger generation. We suggest that a crucial factor that might be responsible for the failure of these falcons to regain their former population level is an inadequate prey base compared to the pre-1960s.

Although the reproductive output was near average by the few successful pairs that had secured pockets of adequate habitat, prey scarcity may have been at the root of the high adult turnover and the eventual abandonment of other sites. Factors that have negatively affected the avian community of the region after the 1960s are changes in agricultural practices and land use, and in particular, the escalating decline in the number and ecological quality of wetlands (Turner at al. 1987). Although we have no data on the prey base in our study area, habitat deterioration can be expected to have resulted in a downward trend in waterfowl, shorebirds, and other Peregrine prey (Dekker 1967).

For example, feral Rock Doves (*Columba livia*), that are a world-wide staple for Peregrines (Ratcliffe 1980; Palmer 1988), became much scarcer than formerly in central Alberta after old-style wooden granaries were destroyed and replaced with metal silos on farm yards (Dekker 1993, 1998). Similarly, Gray Partridges (*Perdix perdix*) and open country passerines have been negatively affected by the use of herbicides, modern grain storage methods, and early fall tilling (D. Dekker, unpublished data). One species that can be expected to have increased is the European Starling (*Sturnus vulgaris*), which is a common prey for male Peregrines in populated regions (Palmer 1988) as well as in our area.

Of special note is the observation that the two remaining Peregrine eyries in the study area are within flying distance (3–10 km) of remnant pockets of natural habitat, characterized by "knob and kettle" topography, formed by glacial deposits, and consisting of a mosaic of wooded hillocks and small ponds. By contrast, Peregrines are absent from river sections where the adjacent uplands have been cultivated right up to valley rim.

The prey base hypothesis is further supported by circumstantial evidence from elsewhere in central Alberta. As of 2006, Peregrines have yet to return to historical cliff breeding sites along other major rivers in agricultural regions of the province (R. Corrigan, unpublished data). By contrast, reintroduced falcons have been nesting in both of the province's large cities since the late 1970s (Holroyd and Banasch 1990). Moreover, at least four Peregrines dispersing from natural cliff sites on the Red Deer River – identified by their band numbers – have become breeding adults in nest boxes on high city buildings and industrial towers in the province (Gordon Court, AFWD, personal communication). Dispersal from natural habitats to urban breeding sites was also noted in recovering Peregrine populations in California (Kaufmann et al. 2004). In Alberta, as elsewhere in North America and Europe, there is mortal competition for urban territories (Tordoff and Redig 2003); VanGeneijgen 2005). While dispersal from urban to natural cliff sites is still rare (Tordoff and Redig 2003), city falcons have been moving out to high structures in rural settings for more than a decade. During the 1990s, offspring of the Edmonton city population, as well as from the Red Deer River population, has spread to four power plants near an Alberta lake, even before nest boxes were placed on the smoke stacks. The main food of these breeding pairs is the Franklin's Gull (*Larus pipixcan*), which is seasonally common and easily caught (Dekker and Taylor 2005).

The abundant prey base of Peregrines nesting on industrial structures and in Alberta cities is in sharp contrast to our study area, where we saw the following signs of apparent food shortages.

(**1**) At site A1, only the males were seen to bring in prey for the nestlings, mostly small birds. Even after absences of 3–4 hours, the females returned without food, although their swollen crop indicated that they had hunted and fed themselves, perhaps at a great distance from the nest site.

(**2**) During a failed nesting attempt at site A2, the adult male, upon returning from long absences (1–3 hours) alternated brooding duties with the female, but he did not always bring food for her ($n = 3$). In another instance, he refused to share a prey item with the aggressively begging and wailing female.

(3) Of 60 prey items brought to VF sites, 73% were small passerines and swallows, 10% waterbirds, 10% Richardson's ground squirrels (*Spermophilus richardsonii*), and 7% unidentified prey. In terms of biomass, the mammalian portion of the diet is substantial. Ground squirrels are locally common and have not been recorded before as prey for Peregrines in central Alberta. However, during years of cyclic abundance, ground squirrels and other rodents accounted for 33% of prey biomass of Peregrines in northern Canada (Court et al. 1988a; Bradley and Oliphant 1991).

We saw no evidence of food shortages at the Prairie Falcon nests. During 64 hours of watching at various sites, the nestlings appeared well fed, often with bulging crops and ignoring newly brought food. Of 28 prey deliveries, 75% were ground squirrels. Rodents were also the main prey of nesting Prairie Falcons in Idaho (Steenhof et al.1999). In a very large sample of food items brought to Idaho eyries only 4% were birds (Holthuijzen 1990). This contrasts with our study in which birds made up 25% of prey items, including five passerines and two Killdeer (*Charadrius vociferus*). Steenhof (1998) did not list shorebirds in the known prey of the Prairie Falcon. However, non-breeding Prairie Falcons commonly hunt sandpipers in Alberta wetlands (Dekker 1982), and Prairie Falcons in Utah preyed on juvenile American Avocets (*Recurvirostra Americana*). Porter and White (1973:52) stated that, where both species of falcon occur together, the Prairie Falcon "assumes the role of generalist while the Peregrine is the specialist." In our study area, the Prairie Falcon could indeed be called a food generalist. By the same token, the Peregrine too is a generalist, as suggested by its use of ground squirrels as well as birds.

CONCLUSION

In contrast to observations by other investigators (Brown and Amadon 1968; Nelson 1969; Walton 1978; Cade 1982), we found the Prairie Falcon no match for the Peregrine in the air. In all incidents of territorial conflict, the Peregrines were clearly the dominant aggressors, whereas the Prairie Falcon's defence was weak and evasive. As in incidents reported from Utah (Nelson 1969; Porter and White 1973), nest site take-overs in our area included at least one case of forced egg abandonment at the expense of the Prairie Falcons. Nevertheless, the Peregrines we observed were not always capable of displacing the Prairie Falcons, which supports the contention that both species are more or less compatible and can exist side by side (Palmer 1988). By way of explanation, we hypothesize that the decisive factor in interspecific combat derives from individual differences in size, dominance, and tenacity, which can either lead to a swift take-over of a nest site or a long standoff. As reported by Dekker (1999) the reintroduced Peregrines were of a smaller size than the original breeding population. Initially, this may have hampered the species' chances of replacing the Prairie Falcon at contested cliff sites.

Furthermore, it was apparent to us that some cliff sites held by Prairie Falcons were easier to defend than others. For instance, two contested sites were situated on partly wooded escarpments that allowed the Prairie Falcons to take shelter and evade attacking Peregrines. By contrast, two other sites that included sheer rock faces were the first to revert back to Peregrines after their 30 years' absence. These findings underscore the theory that falcon eyries, as a function of their specific features, are established by tradition (Ratcliffe 1980; Newton 1979). Traditional site selection is especially important in regions where two competing species of falcon occur sympatrically since it should lead to the prevention or reduction of wasteful and repetitive territorial combat (Cade 1960; Dekker 1967).

In any landscape, the "upper limit to the number of established raptor pairs is set by food or nest sites, whichever is in shorter supply" (Newton 1979:75). A link between Peregrine breeding density and the food supply was reported from various habitats (Court et al. 1988a; Ratcliffe 1980; Nelson 1990; Johnstone 1998; Newton 2003). Based on our observations, we believe that the Peregrine population in our study area was not limited by the availability of nest sites but by its food resource. The fact that the current number of territorial pairs in this study is well below 1960 levels suggests that the carrying capacity of the region, in terms of the Peregrine's prey base, has shrunk. A similar and also untested conclusion was arrived at by White et al. (1988) for the sparse distribution of Peregrines in the Fiji Islands and Vanuata.

Although the Prairie Falcon population is back at its former level, current numbers appear to have reached a plateau as suggested by the intensity of intra-specific territorial disputes. Against conspecifics, the Prairie Falcon is a fierce defender. Unlike Peregrines, which fight along gender lines, Prairie Falcons attack both sexes (Steenhof 1998). In April of 2003 and again in 2004, there were two pairs of Prairies at the same site and the level of hostilities was intense. The resident male repeatedly seized the intruding female, and the intruding male tackled the resident male while the latter was standing on the nest ledge near the brooding female. One day, there were two females disputing the nest ledge while the males were engaged in aerial battle. Eventually, only one of the pairs prevailed. These observations support our conclusion that the Prairie Falcon, unlike the Peregrine in the study area, is not limited by the food resource but by competition for available nest sites.

APPENDIX

Some chronological occupation changes at contested nesting sites.
Site A 1. This north-facing, historical Peregrine cliff was occupied by Prairie Falcons in 1967. In 1999, it was frequented by a pair of reintroduced Peregrines that successfully raised three foster young on the only suitable nesting ledge of this cliff site. The pair returned in 2000 and fledged three young of their own from the same ledge. The following year, the returning Peregrines found the site occupied by a pair of Prairie Falcons. The female Peregrine was new and in her

Figure 12.1. Breeding pairs of Peregrine Falcons along the Red Deer River, Alberta, in 1994–2006.

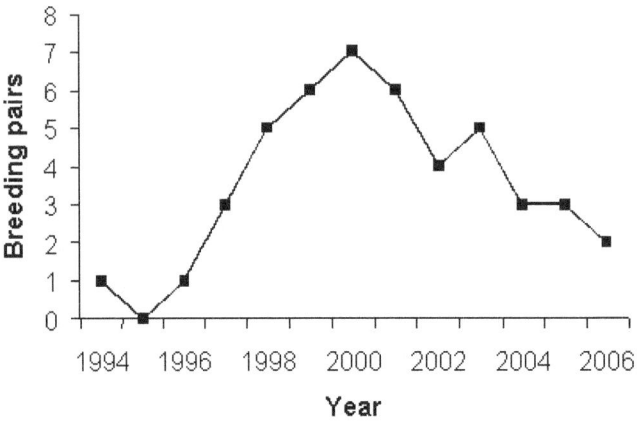

Figure 12.2. Breeding pairs of Peregrine and Prairie Falcons along the Red Deer River, Alberta, 1995–2006.

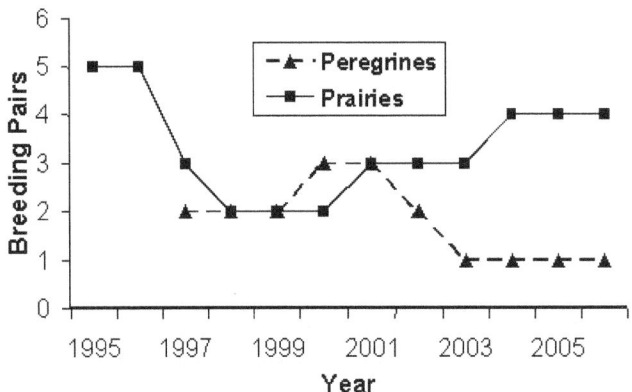

first year, and the pair was unable to dislodge the Prairie Falcons from the superior nest ledge. In that year, and again in 2002, the Peregrines made an unsuccessful nesting attempt on an inferior ledge along the same and largely wooded escarpment, about 100 m away from the Prairie Falcons. In 2003–2006, Prairies again used the superior ledge and we saw no Peregrines there in April and May. Curiously, in late June and early July of 2005, the Prairie Falcons were attacked by a pair of adult Peregrines that appeared to use the inferior ledge as a temporary roost and indicates the existence of non-nesting "floaters" in this population.

Site A2. This prominent 0.5-km-wide south-facing rock wall, historically used by Peregrines, was taken over by Prairie Falcons in 1967 (Dekker 1999). It reverted back to a pair of reintroduced Peregrines in 1997, which held the site until 2001. In June 2001, the adult female Peregrine was found dead on her eggs inside a narrow cavity at the west end of the rock face. The falcon's wing had been partly plucked, and her chest showed two slashes that looked like they had been inflicted by the talons of a raptor, perhaps a Great-Horned Owl (*Bubo virginianus*). Owl predation is a major mortality factor in many Peregrine introduction programs (Tordoff and Redig 2003), but its role in the Red Deer study area is unknown. It is also possible that the Peregrine had suffered these wounds in combat with the pair of Prairie Falcons that was perched beside the nest cavity in which the dead falcon was found two days later. The male Prairie Falcon was actually observed to enter and stay 10–15 minutes inside the cave, enough time to account for the partly plucked wing butt of the dead Peregrine. For the next three breeding seasons, site A2 was occupied by a pair of Prairie Falcons until they were again replaced by Peregrines in 2004. We did not observe the territorial fights that must have preceded this ownership change. However, the circumstances immediately following the switch are noteworthy. On 25 April 2004, an adult male and a first-year female Peregrine had taken over the same nesting cave that had been used by the Prairies. We do not know whether the Peregrines had laid their own eggs or were brooding those of the Prairies. By the end of May it became clear that all nesting attempts had failed. Both the Peregrines and the pair of Prairies were frequenting opposite parts of the escarpment without any apparent hostilities between them. In 2005 and 2006, Prairie Falcons again bred successfully at site A2 and no Peregrines were seen there. (The situation remained unchanged in 2007 and 2008).

Site B1. Its traditional status unknown, this cliff became the scene of prolonged interspecific strife after the Peregrine reintroductions. In 1997, a pair of Peregrines drove off the resident Prairie Falcons, probably forcing them to abandon their eggs or young, but the Peregrines did not nest. In 1998 and 1999, a pair of Prairies used a large nesting cave under an overhang, while the Peregrines laid eggs 35 m away on an open ledge near the top of the same escarpment. After their eggs were predated, probably by a coyote (*Canis latrans*), the Peregrines remained in the area and were subsequently observed to bring food to the Prairie Falcon nestlings under the overhang. The Peregrines appeared to prevent the parent Prairies from entering the nest. During our last visit of the 1999-breeding season there were three fledged Prairie Falcons and two adult Peregrines at the site. In 2000 the Peregrines produced three young of their own under the overhang. Prior to the next breeding season, the overhang collapsed and no young were produced at the site in that year. After a nest box was placed at the top of the cliff and an enclosure built around it to prevent mammalian entry, the site was used by Peregrines for three successive seasons up to and including 2005, but not in 2006.

Hack site No. 1. A curious interspecific conflict occurred at the first hack site in area B soon after the first release of captive-raised juvenile Peregrines in 1992. The displaced pair of Prairie Falcons, whose own young had been removed by AFWD staff, began to deliver food to the recently released Peregrine fledglings. On 15 July, an adult male Peregrine arrived at the site. Alternately attacked by both Prairie Falcons, the Peregrine made vigorous counter-attacks and stayed around for several weeks. He also brought prey items to the hacked fledglings. Identified by his band number, this adult Peregrine was captive-reared and had been cross- fostered into a Prairie Falcon nest in southern Alberta eight years earlier. Cross-fostering of juvenile Peregrines into Prairie Falcon nests was commonly practiced in some western regions of North America (Cade and Burnham 2003). To our knowledge, there is only one record (from Saskatchewan) of a cross-fostered male Peregrine interbreeding with a female Prairie Falcon and raising hybrid young (Oliphant 1991). In addition, on a hack tower in Utah, sometime in the mid-1980s, a male Peregrine mated successfully with a female Prairie Falcon that had been a falconry bird (Jim Enderson, personal communication).

Prairie Falcons, nesting in central Alberta, commonly feed their young on ground squirrels as well as small birds. (Photo: Kay Hodges).

CHAPTER 13

Gyrfalcons and Prairie Falcons hunting city pigeons

Dekker, D. and J. Lange. 2001. Hunting habits and success rates of Gyrfalcons (*Falco rusticolus*) and Prairie Falcons (*Falco mexicanus*) preying on feral pigeons (Rock Doves, *Columba livia*) in Edmonton, Alberta. Canadian Field-Naturalist 115:395-401

Abstract – Three adult female Gyrfalcons (*Falco rusticolus*) and one adult female Prairie Falcon (*Falco mexicanus*), wintering in Edmonton, Alberta, killed 15 and 27 feral pigeons (Rock Doves, *Columba livia*) respectively in 141 and 104 hunts. The Prairie Falcon's hunting success rate (26.0%) was significantly greater than that of the Gyrfalcons (10.6%). Six kills (40%) by Gyrfalcons and 18 kills (67%) by the Prairie Falcon were the result of surprise attacks on sitting pigeons. Open attacks on flying pigeons resulted in nine captures each. The hunting methods of the three Gyrfalcons showed individual differences, and the Gyrfalcon methods differed from those of the Prairie Falcon. In open attacks on airborne flocks of pigeons, the Gyrfalcons swooped up and into the dense flock from below, whereas the Prairie Falcon stooped at individual flock members from above. Both species often resumed hunting immediately or soon after consuming a pigeon. On one occasion, the Prairie Falcon killed four pigeons in 20 minutes of which three were abandoned because of human interference. Besides capturing live prey, a Gyrfalcon took a pigeon from the Prairie Falcon, and another Gyrfalcon scavenged four dead pigeons.

INTRODUCTION

Gyrfalcons (*Falco rusticolus*) are known to take a wide range of avian and mammalian prey including Rock Doves (*Columba livia*) (Palmer 1988; Clum and Cade 1994), but descriptions of their hunting methods are generally based on very few observations and do not include hunting success rates (White and Weeden 1966; Bengtson 1971; Dobler 1989; White and Nelson 1991; Garber et al. 1993). The diet of the Prairie Falcon (*Falco mexicanus*) includes pigeons but consists mainly of small rodents and passerines (Anderson and Squires 1997; Steenhof 1998). The literature contains little information on Prairie Falcons capturing avian prey. Dekker (1982) saw 36 hunts including one kill. Beauvais et al. (1992) recorded 37 hunts with five captures. To the best of our knowledge there is only one published account on the foraging habits of Gyrfalcons wintering in an urban environment (Jenning 1972), and no such information on Prairie Falcons. This paper presents empirical data on hunting methods and success rates of one Prairie Falcon and at least three different Gyrfalcons wintering in Edmonton, Alberta, and preying on pigeons.

Like all big cities, Edmonton harbours a large, year-round population of feral Rock Doves, which commonly attracts raptors. Urban Merlins (*Falco columbarius*), which prey mainly on small birds (James and Smith 1987), occasionally kill pigeons in Edmonton during winter (Lange 1985, and unpublished notes). Peregrine Falcons (*Falco peregrinus*), which are known to hunt pigeons, nest on high buildings in Edmonton, but are absent from mid October to late March (G. Court, Alberta Environment).

Gyrfalcons breed in arctic regions and some migrate south in winter (Godfrey 1986; Clum and Cade 1994). The earliest arrival date for central Alberta is 25 September, and sight records for the city of Edmonton range between November and March (Dekker 1983; Lange 1995). Prairie Falcons are year-round residents in southern Alberta up to the latitude of Red Deer (Godfrey 1986), and they have been sighted in all seasons in the Rocky Mountains up to the latitude of Jasper National Park (Dekker 1982, 1984, and unpublished field notes). Prairie Falcons are transients in north-central Alberta from spring to fall, but there are two nesting record from west and north of Edmonton (Semenchuk 1992). This paper includes a first record of the Prairie Falcon wintering in that city.

STUDY AREA AND METHODS

The study area is centred on the Alberta Grain Loading Terminal (the Granary) situated along the railway corridor that transects the city of Edmonton. The 43 m elevator building includes an annex loading-shed with a flat roof, which is littered with spilled grain, attracting an estimated 600 feral pigeons that roost on the tanks and chutes above the annex and on the window sills of the main building (Figure 13.1).

At the approach of a falcon, most pigeons fly up and draw together in dense flocks that course back and forth over the area or depart. Alerted by such antipredator behaviour, the second author (JL), who works in the rail yards and lives near the Granary, was each year the first to sight falcons. Beginning in mid-December 1998, JL and DD recorded all falcon attacks on pigeons. We defined an attack or hunt as one completed attempt at capturing prey of which the outcome was known (Dekker 1980). A hunt could include one or more swoops or stoops at a pigeon or a flock of pigeons. If the falcon abandoned or interrupted its efforts by perching, its next attack was considered another hunt. Thus, a hunting sequence could include more than one hunt.

The hunts reported here combine those seen by both authors, alone or together. JL's work often placed him in a position to observe falcon activity during the week. In addition, on some days, JL watched from a parked vehicle just south of the Granary. DD watched from the same parking lot on 37 days in 1998–1999 and on seven days in 1999–2000, usually from 11:00 to 16:00. In particular during weekends, one or more local birdwatchers spent varying amounts of time waiting for falcons at the Granary. The pigeon kills they reported to us were not included in our tabulations but contributed to our understanding of the hunting methods used by these falcons. Weather conditions during observations varied and daytime highs ranged from 5° to -30° C. The ground was covered with 10–50 cm of snow.

RESULTS AND DISCUSSION

We saw at least three different adult female Gyrfalcons in the study area. One was mostly white, the others grey variants (Cade et al. 1998). The white "gyr" was lightly spotted on flanks and leg feathers. Dorsally, it was grey-brown with fine barring that gave it a "ladderback" appearance. The crown of its head was light brown unlike other white Gyrfalcons, which have white heads (Brown and Amadon 1968). The white gyr was first seen on 15 January 1995, when it was perched on a staircase railing near the top of the Granary (Lange 1995). It roosted in the same spot over the next six winters although it could be absent for several nights in a row. Use of specific roosting sites with occasional absences of several days was also typical of Gyrfalcons wintering in Washington State (Dobler 1989). The white gyr usually arrived on its roost at dusk and was gone at first light. During the day, it was rarely seen until the winter of 1999–2000 when it often used an antenna on the roof of the Granary as a hunting perch. The last date on which the white gyr was seen each year varied from mid-March to 23 March. This is well past the time most adult Gyrfalcons are believed to return to arctic breeding sites (Platt 1976). At Beaverhills Lake, east of Edmonton, the latest Gyrfalcon sighting, possibly of an immature, was 14 April (Dekker 1983).

Initially, we assumed the white gyr to be a male since it looked smaller, more slender and faster on the wing than a heavily streaked, grey-brown Gyrfalcon (gyr #2) that frequented the study area from 6 December 1998 to mid March 1999. Our assumption was proven wrong on 1 March 2000 when Erhard Pletz captured and banded the white gyr. Its measurements (weight 1780 grams, wing chord 405 mm) placed it well within the range of female Gyrfalcons (Palmer 1988; Clum and Cade 1994).

Gyr #2 was observed during the winter of 1998–1999, but not in 1999–2000. Then, a different grey-brown female falcon (gyr #3) was present from 2 December 1999 to mid March 2000. Ventrally it was less heavily streaked and its feet were a paler yellow than gyr #2. On its left leg, gyr #3 carried a dull grey (as opposed to a new and shiny) metal band. We were unable to read its band number, but a similar-looking grey Gyrfalcon, then a first-year immature, was banded by Erhard Pletz on Edmonton's outskirts on 28 November 1998. Its weight and wing chord were respectively 1580 g and 400 mm.

We saw no interactions between the white gyr and gyr #2, but often between the white gyr and gyr #3. The white gyr was slightly larger than #3 and chased it off aggressively. In addition to the three falcons described above, we saw two or three other Gyrfalcons near the granary (Court 1999), but they were not in view long enough to allow for accurate details as to colour, age group, or sex.

The first Edmonton record for the Prairie Falcon was 24 December 1995 (Lange, unpublished notes). Its yellow feet, large size and prominent dark axillars identified it as an adult female (Wheeler and Clark 1995). In this paper, we refer to it as "the" Prairie Falcon although we have no proof that all sightings involved the same individual. During the winters of 1996–1998, JL and several associate observers had seen the Prairie Falcon attack and capture at least eight pigeons at the Granary, but no detailed tally of hunts and kills was kept until the start of this study. During the winter of 1998–1999, the Prairie Falcon was present from 5 December to 26 February. In 1999–2000, it was first sighted on 2 December but not after the end of December.

Gyrfalcon hunting methods. Gyrfalcons commonly hunt very low over the ground and seize avian prey just after it flushes. The target may be first spotted from an elevated perch (White and Weeden 1966; Bengtson 1971; Palmer 1988; Clum and Cade 1994). A similar strategy of surprise was also employed by the Gyrfalcons in this study. From a high perch on the roof of the Granary, the falcons launched attacks on pigeons seeking food or grit along the railway or on snow-free parking lots in the neighbourhood. The hunt could be partially obscured from our view behind buildings so that its outcome could not be determined unless the falcon came back into view, either with or without prey. The falcons directed similar surprise attacks on pigeons sitting on the flat roofs of adjacent, lower buildings. The above methods resulted in two captures (Table 13.1). Three similar kills were reported to us by associate observers. In addition, gyr #2 was seen to drop down from its high granary perch to seize a pigeon just as it flushed from the lower roof of the annex.

Surprise was also the initial strategy of Gyrfalcons approaching the Granary from afar. Flying between adjacent buildings or along the railway, their sudden arrival caused the pigeons to flush in a panic from the annex or the front wall of the Granary. Typically, the falcon shot upwards through the dense flock. In four instances, it seized a pigeon at once. Associate observers reported two similar captures. In a variation of these surprise tactics, gyr #3 started its attack from a perch on the side of the Granary. Flying around the corner of the building, it seized a pigeon from a flock that flushed from the lower annex.

If the falcon's initial pass failed, it might leave the area or perch on the roof of the Granary. However, sometimes it continued its attacks in an open and deceptively casual manner, flying back and forth near the careening pigeons. After gaining a height advantage, the falcon suddenly descended in a burst of speed with beating wings and, carried by its momentum, swooped upwards into and through the flock. Twisting aside or "standing on its tail" and practically stalling at the apex of its swoop, the falcon attempted to seize the nearest pigeon. If it failed to make a capture, it turned back down and might repeat its upward swoops up to six times in a row. Alternately descending and ascending, the falcon typically attacked "like a pendulum." These open attacks resulted in eight captures, and visiting birdwatchers reported another eight similar captures, all by the grey falcons.

The white gyr rarely attacked flocks in the above open style. Instead, it launched direct pursuits of lone pigeons or small groups passing by the Granary, often well over 100 m distant. Starting from its high perch on the antenna, the white gyr quickly overhauled the pigeons, which dodged the attack at the very last moment. The falcon seldom made a second try at the same target. Despite its impressive speed, the white gyr was usually unsuccessful. On 17 February 2000, in a three-hour period, it made at least 12 fruitless attacks on single pigeons and flocks, alternated with spells of perching. On 20 February, over two hours, it made about 20 futile hunts. We saw only one kill by the white gyr. Here too, surprise may have played a role. Breaking off its attack on a flock, the gyr flew about 50 m in pursuit of a single pigeon that had left the flock. Failing to dodge, the pigeon was seized from behind. Direct pursuits of single pigeons flying by the granary were not made by Gyr # 2 and rarely by Gyr #3, but always without success.

Prairie Falcon hunting methods. Prairie Falcons are known to take a wide variety of prey, including pigeons (Palmer 1988). They often soar to great altitudes and descend in long glides to surprise small mammals and birds (Dekker 1982). During winter they generally fly low over open ground (Anderson and Squires 1997; Steenhof 1998). Surprise was also a prominent strategy in 67% of the kills made by the Prairie Falcon observed in this study. Its most common method was a sudden approach to the Granary. We rarely spotted the falcon's arrival before the pigeons flushed in alarm. Swooping through the flock, it could be immediately successful (Table 13.1). If it failed, it often perched on the Granary. Unlike the Gyrfalcons, which chose higher resting spots on the roof or the antenna, the Prairie Falcon always selected a metal bracket on the front wall of the Granary well below the roof but still about 25 m above the annex. After the pigeons had settled back onto the flat roof and pipes below, the falcon suddenly dropped off its perch and stooped down perpendicularly. While the pigeons flared away from the building, the falcon shot through them and might at once emerge carrying its prey. In a similar attack, when few pigeons were present, the stooping falcon passed in between the pipes above the annex, levelled off just over the flat roof and seized a sitting pigeon from behind, carrying it along without pause.

If its surprise tactics failed, the Prairie Falcon either: (1) left the area; (2) returned to its bracket perch and tried again 5–25 minutes later; or (3) made open attacks on flocks that had already been flushed. Flying back and forth, or sailing and soaring, the falcon stayed high above the careening pigeons, until it suddenly stooped down with rigid wings, either fully extended or tucked in close to the tail. It usually aimed its attack at a specific pigeon on the outside of the flock. If the target dodged, the falcon pulled up and stooped at other pigeons, sometimes five or six in a row. Some hunting sequences lasted 5–10 minutes with the falcon sailing away some distance and returning to attack again. Meanwhile some pigeons might settle back onto the Granary. Others had stayed put, crouching on windowsills or on the pipes and tanks above the annex. The falcon flushed these pigeons by stooping at them, but it seldom chased them for more than 20 m.

On 4 January 1999, the Prairie Falcon made a series of four kills in about 20 minutes. The first three pigeons were taken from flocks at the falcon's first pass and carried to a nearby snow-covered parking lot. Their carcasses were abandoned a few minutes later, probably because of the approach of vehicles. Each time, the falcon flew directly back to the Granary and seized its next prey from the flushing flock. After its third kill, the falcon perched briefly on a pole by the parking lot. It then returned to the Granary and caught a fourth pigeon, carrying it out of sight. Searching the parking lot and by following a trail of blood drops, associate observer Gordon Court and DD located the falcon's three previous kills lying in 30 cm of soft snow.

Food habits, scavenging, and kleptoparasitism. On several days, a Gyrfalcon that had just consumed a pigeon began attacking pigeons again immediately afterwards. The Prairie Falcon resumed its attacks on pigeons about 25 minutes after it had captured and eaten one. On two occasions, it killed and fed upon a second pigeon within two hours from the first. As Rock Doves weigh 250–400 g, the falcons in this study apparently killed well in excess of published daily food requirements, which are assumed to be 250 grams for Gyrfalcons and 170 grams for Prairie Falcons (Craighead and Craighead 1956; Palmer 1988).

All captures, including those reported by associates, occurred between 10:30 and 16:30. Although we saw no hunting or feeding activity before 10:00, that possibility remains to be investigated more closely. In this regard it may be significant that the white gyr, on the day it was trapped, came down to the lure pigeon at first light. The previous evening it had arrived at its roost with a bulging crop as proof that it had eaten well (E. Pletz, personal communication).

Both species of falcon carried their freshly caught pigeons to open terrain such as the parking lot adjacent to the Granary. In flight, they bent down to bite the pigeon in the neck, probably to kill it. We did not see them pluck prey on poles or buildings. The Gyrfalcons did not avoid snow of 20–40 cm deep, but the Prairie Falcon usually selected bare ground along the railway. The falcons sometimes carried their prey to the Municipal Airport, about 1 km away, where the runways were mostly clear of snow.

We saw little overt aggression between the Gyrfalcons and the Prairie Falcon. Gyr #2 occasionally chased the Prairie Falcon. At other times, both perched on the Granary at the same time. Opportunistic klepto-parasitism might be common but we saw only one such incident. On 13 February 1999, the Prairie Falcon caught a pigeon at the Granary, while gyr #2 was perched on the roof. It took off at once in pursuit. After the loss of its prey, the Prairie Falcon turned back to the Granary, flushed the flock again, and caught another pigeon at its first pass. We did not see any intraspecific klepto-parasitism between the Gyrfalcons.

Gyrfalcons are known to feed on carrion in their northern breeding grounds (Dementiev 1960; Palmer 1988; Cade et al. 1998). On 19 February 2000, after eight unsuccessful attacks on flocks of pigeons, gyr #3 landed on the flat roof of the annex, walked to the remains of a dead pigeon, and flew off with it. On 21 February, the same falcon retrieved two dead pigeons in about 40 minutes. As proof that it had eaten from the first carcass, the gyr showed a bulging crop. On 22 February, it again took a dead pigeon from the annex. Clutching the carcass in its feet, the gyr soared over the area and drifted away.

On 24 February, to investigate the causes of death, Gordon Court removed eight pigeon carcasses from the annex. Apparently, their death had been accidental, caused by hitting roof fans. However, it is possible that some had died as a result of wounds caused by falcon attacks. On one occasion, we watched gyr #2 swoop upwards through a flock with the result that one pigeon dropped out and fell like a stone in between the machinery above the annex. The falcon made no effort to retrieve it. It is possible that other pigeons were wounded and escaped to die later. While being carried away by a Gyrfalcon, one pigeon struggled free and flew off, but the gyr did not pursue and recapture it. In one incident we saw the Prairie Falcon strike a flying pigeon a mortal blow that caused it to stop beating its wings and make a tumble. The falcon instantly doubled back and caught the pigeon before it had fallen more than 5 m. In all other captures the Prairie Falcon directly seized the prey in its feet.

Contrary to reports that Gyrfalcons commonly strike prey in the air, so that they drop to the ground (Dementiev 1960; Clum and Cade 1994), the Gyrfalcons in this study bound to their prey. In close but unsuccessful attacks, the pigeons might lose a puff of feathers, or the falcons trailed feathers from their claws even though all pigeons flew on. Such damage was likely the result of misdirected attempts at seizing prey. Loose feathering in pigeons may be an anti-predator adaptation. Seizing prey, as opposed to striking them down, is also the most common capture method of the Peregrine Falcon (Dekker 1999).

The avian prey of Gyrfalcons consists mainly of grouse and waterfowl (Brown and Amadon 1968; Palmer 1988; Clum and Cade 1994), which are less manoeuvrable in flight than pigeons. Jenning (1972) noted that a Gyrfalcon wintering in Stockholm, Sweden, ignored pigeons. Bent (1937–1961:2) cited three eyewitness accounts to the effect that Gyrfalcons often tried but "never succeeded in capturing a pigeon.... which were more than a match for them." This contention is supported by some of our observations, such as the ineffectiveness in pursuit shown by the white Gyrfalcon. However, the overall success of the Edmonton Gyrfalcons was 10.9%, which is close to the mean (12.7%) success rate of 13 studies of wintering and migrating Peregrine Falcons (Roalkvam 1985). The gyrs in this study made effective use of surprise, which concurs with known strategies. However, their persistence in attacks on flying flocks of pigeons and the "pendulum-style" swoops reported here have not been described before.

It is interesting to note that the Prairie Falcon seemed capable of capturing its prey at will. The pigeon's habit of flying up and drawing together at the sight of raptors is considered to be an anti-predator strategy that benefits individual

A diagnostic feature of the Prairie Falcon are the dark underwing coverts that distinguish it from other large falcons. It often soars at a great altitude and launches long stoops at ground-dwelling birds and rodents. (Photo: Gerald Romanchuk)

survival. However, the dense flocks, careening back and forth, created an easy opportunity for both the Gyrfalcons and the Prairie Falcon.

A comparison of hunting success rates between different species of raptors means little unless they hunt the same kind of prey, at the same locality, and during the same season. These criteria are met in this study and support our conclusion that the success rate of the Prairie Falcon in capturing pigeons was significantly greater than that of the Gyrfalcons. This observation fits the theory that a falcon pursuing prey in the air should have a body mass close to that of the prey in order to match the speed and manoeuvrability of that prey (Anderson and Norberg 1981; Cade 1982).

Addendum. During the very mild winter of 2000–2001, the white Gyrfalcon failed to return to the Edmonton Granary, and sightings of grey Gyrs were rare in the city. Based on plumage characteristics, a Prairie Falcon seen at the Granary on 13 March proved to be a different individual than the bird seen in 1998–2000. Yet, it used exactly the same perch on the Granary and attacked the pigeons in similar style. It appeared to be less accomplished, however, as all of its 18 hunts, made over a 3-hour period, were unsuccessful. During the winter of 2001–2002, a Prairie Falcon was again frequently seen attacking pigeons at the granary, but Gyrfalcon sightings were not reported.

Table 13.1: Captures of feral pigeons by Gyrfalcons (GP) and Prairie Falcons (PF) wintering in Edmonton, Alberta.

	GF	PF
Low surprise attack on pigeon(s) sitting on the ground	1	–
Surprise attack on pigeon(s) sitting on flat-roofed buildings	1	–
Vertical stoop from high perch on building at pigeons sitting 25m lower down	–	10
Sudden attack on flock flushed from building, immediately successful	4	8
Repeated attacks on free-flying flocks, prey seized from flock	8	8
Pursuit of lone pigeon leaving flocks under attack	1	1
Totals	15	27
Total number of hunts	141	104
Success rate	10.6%	26.0% *

*$G = 6.82$, $P < 0.01$ (Sokal and Rohlf 1969)

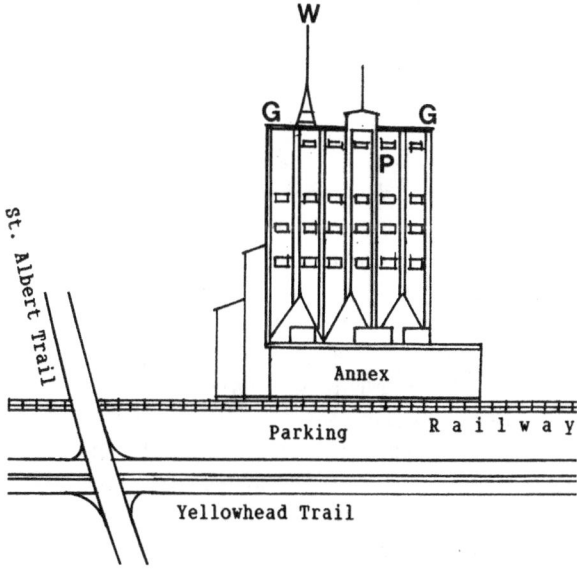

Figure 13.1. Location sketch of the Alberta Grain Terminal near the intersection of Yellowhead Trail and St. Albert Trail in northwest Edmonton, Alberta. The habitual perches of the white Gyrfalcon, the two grey Gyrfalcons, and the Prairie Falcon are indicated, respectively, by the symbols W, G, and P.

CHAPTER 14

Gyrfalcons and Bald Eagles hunting wintering Mallards

Dekker, D. and G. Court. 2003. Gyrfalcon predation on Mallards and the interaction of Bald Eagles wintering in central Alberta. The Journal of Raptor Research 37:161–163.

Abstract – Mallards (*Anas platyrhynchos*) wintering on an unfrozen stretch of river downstream from the city of Edmonton, Alberta, commuted daily to stubble fields and cattle feedlots, where they were attacked by Gyrfalcons (*Falco rusticolus*). Seventy hunts resulted in 16 known and four probable kills. The majority (77%) of Mallards chased by Gyrfalcons escaped capture by taking cover in low vegetation, under bushes, or on roadsides. Prey were seized on the ground or in the air and borne down. The Gyrfalcons used two main hunting methods: (1) low surprise attacks on feeding ducks; and (2) long-range intercepts of high-flying ducks. Respective success rates for these hunting methods were 12% and 40%. Bald Eagles (*Haliaeetus leucocephalus*) flushed and chased Mallards over stubble fields, rarely succeeding, but they habitually pirated duck kills from the Gyrfalcons.

INTRODUCTION

The food habits of Gyrfalcons (*Falco rusticolus*) have been studied mainly by the collection and identification of prey remains (Palmer 1988; Poole and Boag 1988; Cade et al. 1998). Field observations of Gyrfalcons capturing prey are few and anecdotal (White and Weeden 1966; Bengtson 1971; Dobler 1989; Garber et al. 1993). The largest data sets of hunts and kills by wintering Gyrfalcons originate from urban study areas. Jenning (1972) and Dekker and Lange (2001), respectively, detailed the capture of 17 Mallards (*Anas platyrhynchos*) and 15 Rock Doves (*Columba livia*) in Stockholm and Edmonton. We present a comparable sample of Mallard kills by Gyrfalcons wintering in a rural region of Alberta. Bald Eagles (*Haliaeetus leucocephalus*) usually forage over water, and their kleptoparasitic habits are well known, particularly at the expense of the Peregrine Falcon (*Falco peregrinus*) (Anderson and DeBruyn 1979; Dekker

1985, 1987, 1995). In this paper, we describe observations of eagles hunting ducks over land and pirating prey from Gyrfalcons.

STUDY AREA AND METHODS

The latitude of the study area is 53°N and the climate is cold continental. The North Saskatchewan River, that flows through Edmonton, Alberta, is frozen from November to May except for a stretch of roughly 10 km downstream from the city. The open water attracts >1000 Mallards that stay all winter. In fall, they make daily feeding flights of 2–10 km over gently undulating agricultural plains. After the ground is covered with >10 cm of snow, the ducks abandon the stubble fields and congregate at farms where cattle are being fed with grains or silage. Gyrfalcons are migrants and winter residents in central Alberta. Earliest and last records are 25 September and 14 April, although the majority of sightings date from November to mid March (Dekker 1983; Court 1999). Bald Eagles winter in south-central Alberta where lakes or rivers remain ice free (Godfrey 1986).

To study the interaction between ducks and Gyrfalcons we used three principal methods: (1) Sitting in a parked vehicle, we monitored ducks feeding on fields or in farmyards for sudden alarm behaviour caused by the arrival of a predator; (2) From a high viewpoint, we frequently scanned the skies through binoculars, waiting for Mallards to fly inland from the river; and (3) Gyrfalcons, either in flight or perched, were watched in anticipation that they might initiate the hunt for prey. Over three winters, 1999–2002, we visited the study area on 95 days (3–5 hours/day) and sighted one or more Gyrfalcons on 56 days.

A "hunt" was defined as one completed attempt by a Gyrfalcon at capturing a duck of which the outcome became known to us (Dekker 1980). The term "kill" means that we saw the Gyrfalcon pursue and seize a duck, or that we located the falcon on its prey shortly after the hunt. A "probable kill" means that a Gyrfalcon, in close pursuit of a duck, chased it down to the ground and disappeared from view behind trees or sloping terrain in the distance. Additional clues were provided by Bald Eagles that flew to and descended in the same locality of the suspected Gyrfalcon kill.

RESULTS AND DISCUSSION

Gyrfalcons. During each winter, 1999–2002, we sighted at least four different Gyrfalcons, ranging from very dark immatures to partly white adults (Court 1999; Dekker and Lange 2001). Some recognizable individuals stayed in the study area for a period of several days or weeks, others were sighted only once.

Gyrfalcons are known to hunt low over the ground and seize avian prey just after it flushes. The target may be first spotted from an elevated perch (White and Weeden 1966; Bengtson 1971; Palmer 1988; Clum and Cade 1994). A similar strategy of surprise was employed by the Gyrfalcons in this study. Their sudden arrival, flying low over the ground, caused foraging ducks to fly up. After selecting an individual target, the Gyrfalcon chased it on an erratic, twisting course. If the hunt failed, the falcon perched on a post or tree and presently launched another hunt. We saw a total of 42 low surprise hunts of which five resulted in a kill (Table 15.1). In four additional instances, we just missed seeing the attack and discovered the falcon on a fresh kill. Six successful low surprise hunts were reported to us by associate observers.

A very different strategy was employed by Gyrfalcons that flew out to meet ducks in flight on their way to feeding sites at altitudes of 50–300 m and still >2 km away. Leaving a perch, the Gyrfalcon began a long climb, gradually increasing speed. At the falcon's approach, the ducks turned away and split up. In the terminal stage of the hunt, the Gyrfalcon selected a single duck that appeared to realize its vulnerability and descended in an attempt to reach cover. Twenty-five of these open hunts resulted in ten confirmed and three probable kills (Table 14.1).

Most pursuits ended in a close tail-chase. In the two exceptions the Gyrfalcon flew 20–100 m higher than the duck and stooped at it nearly vertically. In both cases, the duck dodged the stoop and landed, followed by the falcon. One was a confirmed kill, the other a very probable kill. We saw no evidence that the falcons struck and knocked down their prey. In all observed captures, the prey was seized in mid-air or on the ground.

The majority (77%) of ducks that were closely pursued escaped into low vegetation protruding from the snow, or under bushes or roadside trees. In two instances, the falcon landed on the road shoulder and, in a futile attempt to flush the duck, walked into the snow-drifted ditch under the trees. Eight ducks landed on the weedy shoulders or medians (10–15 m wide) of divided highways, and two of these ducks were seized by the falcon while heavy traffic passed by. In other cases, the falcon repeatedly swooped at the duck but failed to grab it. One drake that landed on the pavement of a highway defended itself by lunging with gaping bill at the swooping falcon. After several failed passes, the Gyrfalcon perched in a tree, but immediately started in pursuit again when the drake took off, until it eventually reach cover. Bird bander Erhard Pletz (personal communication) saw a Gyrfalcon pursue a duck until it went down on an open, snow-covered field. The falcon swooped at the duck but did not seize it and landed 1–2 m away. Similar stand-offs between Gyrfalcons and Mallards were described by Jenning (1972). Gyrfalcons are also reluctant to grab a tethered decoy pigeon if it fails to fly up (Dekker and Lange 2002). Apparently, they prefer to grab their prey from behind while it is fleeing from them.

(There is a parallel here with cursorial hunters such as wolves (*Canis lupus*), which are stimulated to attack moose that run from them, but hesitate if the animal makes a stand (Mech 1966). A similar standoff relationship between wolves and elk was seen by Smith et al (2007) in Yellowstone National Park).

A realistic assessment of the hunting success of the Gyrfalcons in this study is difficult. Based on confirmed kills, the success rate is 22.9%. The addition of probable kills brings the rate up to 28.6%. This is more than twice as high as the 10.6% success rate in 141 hunts by Gyrfalcons preying on Rock Doves (Dekker and Lange 2001). Mallards are less manoeuvrable than pigeons and usually escape from falcons by plunging into water. In this study, the Gyrfalcons outflew and forced down any Mallard they pursued with persistence. Although the long-range, high altitude interception of flocks of ducks by Gyrfalcons has not been described in the literature, direct climbing attacks on other bird prey have been reported before and the Gyrfalcon's capacity in this regard is well known to falconers (Cade 1982). An identical mode of hunting, launched from a perch and aimed at the interception of high-flying ducks, was employed by some male Peregrines (*F. p. pealei*) wintering in British Columbia (Dekker 1995, 1999).

Although some Gyrfalcons observed in the study area were seen hunting Rock Doves or found feeding on Gray Partridges (*Perdrix perdrix*), their main prey appeared to be the locally wintering Mallards. Probably in response to predation risk, the ducks delayed their feeding flights until evening. During fall, the flocks routinely travelled to the stubble fields near noon. However, after they had been attacked often, they did not leave the river until near sundown. For instance, on 18 February 2002 the first flock (20–25 ducks) departed at 15:00. As soon as they were met by a Gyrfalcon, the flock turned back and escaped into the water. The falcon remained in the area, often switching perches. Two hours later, the ducks suddenly lifted off again in multiple flocks totalling hundreds of birds. At the approach of the Gyrfalcon, only the flock under direct attack returned to the river, pursued by the falcon, while other flocks pressed on to the feeding area about 8 km away.

Of nine Mallard kills of which the sex was known, seven were drakes and two hens. In two instances, a Gyrfalcon ate only part of a drake and returned to the remains the following morning. Two other falcons consumed all flesh from the carcass in 35–45 minutes, leaving only the head, the pelvis, and the lower portion of the legs. As reported by a farmer who shot a number of raiding Mallards, by late winter these ducks were emaciated.

Bald Eagles. During the freeze-up period, the last of the Bald Eagles migrating through central Alberta actively hunt ducks in water holes of lakes (Dekker 1984c, 1985). However, the food habits of the Bald Eagles wintering in the study area have not been studied in detail. Each year, we counted one or more eagles perched on trees along the open stretch of the river. One adult eagle was seen to lift a Mallard from the water. In late fall, the eagles also perched on trees over-looking the stubble fields where Mallards congregated. Some eagles actively hunted the ducks in fast, low surprise attacks. However, all of 12 chases of ducks that flushed just ahead of eagles were unsuccessful, but one eagle doubled back and pounced on a drake that had turned back and landed again. Associate observers reported three additional captures of Mallards of which one was taken in flight (birdwatcher Fred Whiley, personal communication).

After the fields were covered in snow and the ducks commuted daily to cattle feedlots, the eagles left the river and perched on trees in view of the feeding ducks. We suspect that they were also spying on other raptors. We never saw Gyrfalcons attack ducks on farms where eagles sat waiting on prominent perches. However, if falcons hunted in the distance, it was common to see eagles fly into that direction, apparently searching the fields. We saw four eagles rob Gyrfalcons, which released their ducks at once. Two additional cases of kleptoparasitism were reported by associate observers. The Gyrfalcons did not defend their prey against the eagles. Other raptors seen to hunt the ducks were Northern Goshawk (*Accipiter gentilis*) (six low attacks), Prairie Falcon (*Falco mexicanus*) (one long chase), and Snowy Owl (*Nyctea scandiaca*) (two low attacks). One owl, after waiting for a Gyrfalcon to finish eating, scavenged the remains of the duck.

Table 14.1. Hunting methods of Gyrfalcons preying on Mallards

Hunting method	# Hunts	# Kills	# Probable kills
Stealth attack on sitting ducks	42	5	–
High, open attack on flying ducks	25	10	3
Unknown approach	3	1	1
Totals	70	16	4

Above: Adult Gyrfalcon in typical grey plumage, photographed by Miechel Tabak at Boundary Bay in British Columbia. There, it habitually robbed Peregrines of their Dunlin prey. Gyrfalcons as well as Peregrines were kleptoparasitized by Bald Eagles. The photo on the opposite page shows a very dark adult Gyrfalcon that wintered near Edmonton, Alberta, in 2002 and 2003. (Photo: Gerald Romanchuk).

CHAPTER 15

Herftstrek en jachtgewoontes van de Slechtvalk in Friesland.

Slechtvalk predatie op overwinterende Bonte Strandlopers in vergelijking tot de westkust van Canada

Dekker, D. & A. Ferwerda. 2008. Slechtvalken in Noard-Fryslan Butendyks. Twirre19(1):2–10.

De sterke terugkeer van de Slechtvalk heeft recent geleid tot onderzoek naar de indirecte gevolgen van de toegenomen predatiedruk op het gedrag en gewicht van prooivogels als steltlopers. Er is echter weinig studie gemaakt van in Nederland overwinterende Slechtvalken die op strandlopers jagen. Dit gemis aan gegevens leidde ertoe dat wadvogelecoloog Theunis Piersma de Canadese valkenexpert Dick Dekker uitnodigde om intensief naar Slechtvalken te komen kijken in Nederland. Piersma raadde Noard-Fryslân Bûtendyks aan als het meest geschikte terrein. In de periode van 30 oktober tot 4 december, 2006 en 2007, heeft Dekker daar 59 dagen valken geobserveerd, in samenwerking met terreinopzichter Albert Ferwerda.

INLEIDING

Na hun kritieke, door landbouwgifstoffen veroorzaakte achteruitgang in de periode 1960–1970, zijn Slechtvalken in alle vroegere broedgebieden weer toegenomen (Hustings & Van Winden 1998a; Cade & Burnham 2003). In 2006 waren er zelfs 22 nestparen in Nederland, die gebruik maakten van hoge gebouwen en elektriciteitsmasten (Van Geneijgen 2006). Daarnaast zijn ook de aantallen van doortrekkende en overwinterende valken sterk gestegen (Ouweneel 1995; Koks 1998; Hustings & Van Winden1998b; Marcus 2005).

Gebieden die door hun vogelrijkdom uitermate geschikt zijn als winterterritoria voor noordelijke Slechtvalken liggen voornamelijk in de kustprovincies van Groningen tot aan Zeeland. Bekende voorkeursplekken zijn onder andere het Lauwersmeergebied, de Waddeneilanden en de Greidhoeke van Fryslân. Gegevens over aantallen en verzamelde prooiresten zijn te vinden in Beemster (1993), Smit (2000), Brandenburg & Riemersma (2002), en Oosterhuis & Kok

(2003). Wokke (2002) presenteert naast gegevens over prooiresten ook anekdotische informatie over de jachtwijze van de Slechtvalk in een Noord-Hollands polderlandschap. Maar er is weinig gepubliceerd over de predatie op watervogels, iets waar de Canadese veldstudies op toegespitst waren.

In Canada, over een tijdperk van 45 jaar, zag Dekker Slechtvalken enkele duizenden keren jagen, verdeeld over alle seizoenen en in zes verschillende landschappen. Er werden 460 prooien gevangen: 50% steltlopers, 21% eenden, 17% meeuwen en 12% overige soorten. In deze laatste groep zaten slechts drie Rotsduiven. Elders – in bewoonde streken – vormen duiven het hoofdvoedsel van de Slechtvalk (Ratcliffe 1980; Frank 1994; Geneijgen & Van Dijk 1997; Treleaven 1998; Van Dijk 2000). De in Canada opgedane ervaring was van grote waarde om inzicht te verkrijgen in de dynamiek tussen Slechtvalken en watervogels in Noard-Fryslân Bûtendyks. In dit artikel worden vergelijkingen getrokken tussen beide gebieden.

Recent is veel aandacht besteed aan de vraag in hoeverre de ecologie van overwinterende steltlopers wordt beïnvloed door roofvogelpredatie (Piersma *et al.* 2003; Ydenberg *et al.* 2004). De grondbeginselen zijn welbekend. Om naderend gevaar bijtijds te ontdekken, zoeken wadvogels elkaars gezelschap, volgens het principe dat twee paar ogen meer zien dan een paar, en geven zij de voorkeur aan een open leefgebied (Ydenberg & Prins 1984). Als het wad onderstroomt, verhuizen zij naar hoogwatervluchtplaatsen die aan dezelfde voorwaarden moeten voldoen. Een van de onderzoeksvragen die wij ons stelden was: Waar verblijven de Bonte Strandlopers die buitendijks overwinteren tijdens hoogtij? Gaan ze in Fryslan dan ook urenlang boven de zee rondvliegen, zoals hun soortgenoten aan de westkust van Canada? Daar doen ze het blijkbaar om te ontkomen aan het overrompelingsgevaar van Slechtvalken en Smellekens.

STUDIEGEBIED EN WERKWIJZE

Noard-Fryslân Bûtendyks is een 23 km lange kuststrook gelegen tussen Zwarte Haan en Holwerd, in breedte variërend van 0,3 tot 3,0 km, die bestaat uit slikvelden, kwelders en zomerpolders. Het 4.100 ha grote gebied ligt besloten tussen de zeedijk en de Waddenzee en is vanaf 1996 in toenemende mate deels in eigendom en beheer van It Fryske Gea. Het terrein is doorsneden met afwateringssloten en 2–3 m hoge zomerdijken. De meest prominente hoogtes worden gevormd door een aantal ronde dobben van 6–8 m hoogte met een centrale kom van zoetwater, die dient als drinkplaats voor het vee dat van april tot oktober wordt ingezet om de begroeiing kort te houden. Voor een beschrijving van de buitendijkse vegetatie en de kwelderwerken wordt verwezen naar Hosper & De Vlas (1994) en Dijkema *et al.* (1996); voor de broedvogelpopulatie wordt verwezen naar Bos *et al.* (2007). De overwinterende kustvogels worden periodiek geteld door de Wadvogelwerkgroep van de FFF. Zie bijvoorbeeld Engelmoer (2001) voor een gedetailleerd rapport. De aantallen ganzen, eenden, steltlopers en roofvogels die in het poldergedeelte van Noard-Fryslân Bûtendyks overwinteren, variëren van jaar tot jaar en zelfs van dag tot dag. Wij noteerden opmerkelijke aantalsschommelingen van eenden en steltlopers alleen wanneer die betrekking konden hebben op de aan- of afwezigheid van Slechtvalken.

In 2006 verbleef Dekker van 1 november tot 24 november dagelijks in het veld; in 2007 van 30 oktober tot 4 december (met uitzondering van 22 november). Het aantal dagen was respectievelijk 24 en 35. Op overeenkomstige datums van de twee jaren is de doorgebrachte tijd, variërend van een tot acht uur per dag, hier samengevoegd. In 2006 spendeerde hij 34 van de 113 uur aan de westkant van Zwarte Haan, waar de kwelder en de rijsdammen eindigen en de zeedijk een open blik biedt op het aangrenzende wad. De resterende 79 uur en alle 181 uren van 2007 bracht hij door ter hoogte van Hallum en Ferwert, voornamelijk in het Noarderleech en de Bokkepollen, met een paar korte excursies naar de Bildtpollen of Ferwerderadeels Buitendijks. De zomerpolders en kwelders zijn hier 2–3 km breed. In 2007, werd Dekker vaak vergezeld door Ferwerda.

Om valken op te sporen liepen wij via de zomerdijken naar de zeekant van de kwelder. Ook zaten we vele uren lang in een op de zeedijk geparkeerde auto, op punten die een weids uitzicht boden over Noard-Fryslân Bûtendyks waar we met verrekijker en telescoop het terrein afzochten. Slechtvalken die op palen zaten of gewoon op de grond werden voor kortere of langere tijd in het oog gehouden in de hoop dat ze gingen jagen. Vliegende valken werden zo lang mogelijk met de kijker gevolgd.

Een tweede methode was het langdurig observeren van groepen prooivogels, zoals steltlopers, eenden en Spreeuwen. Paniekreacties, in het bijzonder het plotseling en massaal opvliegen, attendeerden ons soms op de aanwezigheid van roofvogels. Hoewel geduld en oplettendheid cruciale vereisten waren voor dit soort onderzoek, dankten wij een aantal van de waarnemingen ook gewoon aan het toeval.

Voor zover mogelijk verdeelden wij de valken in verschillende categorieën. Naast andere verschillen in verenkleed, zijn volwassen Slechtvalken herkenbaar aan hun blauwgrijze rugkleur. Onvolwassen, eerstejaars vogels, zijn overwegend bruin. Mannetjes hebben slankere vleugels dan de vrouwtjes en zijn ongeveer een derde kleiner (Ratcliffe 1980). Onder veldcondities zijn kleur en afmetingen echter relatieve waarden en lang niet altijd met zekerheid vast te stellen. Om die reden zijn wij uiterst kritisch geweest in onze beoordeling. Observaties van korte duur die geen honderd procent betrouwbaarheid konden geven werden niet meegeteld, evenmin als vermoedelijke dubbeltellingen.

Hoewel wij primair geïnteresseerd waren in de voedselgewoontes van de plaatselijke valken hebben wij geen prooiresten verzameld. De reden daarvan is tweeledig. Op de eerste plaats blijft er van kleine prooien, zoals strandlopers, die door een Slechtvalk geplukt worden heel weinig over, vaak niet meer dan veertjes. Bovendien worden zulke lichtgewicht prooien, alvorens ze gegeten worden, meestal ver weggedragen. En op de tweede plaats, kan men van prooiresten niet zeker zijn van de doodsoorzaak. De dader laat geen 'visite-kaartje' achter en er zijn andere roofvogels die op dezelfde prooisoorten jagen als de Slechtvalk, zoals Havik, Sperwer, Smelleken en Blauwe kiekendief. Daarnaast zijn ook meeuwen, vooral de Grote mantelmeeuw, in staat een steltloper of eend bij verrassing te overmeesteren.

In dit artikel beschrijven wij het jachtgedrag en prooikeuze van de Slechtvalk op basis van directe observaties. De termen 'aanval' of 'jachtvlucht' betekenen dat de valk een doorgevoerde poging deed een prooi te pakken, die dan probeerde te ontsnappen door snelle wendingen te maken of dekking te zoeken in ruigte of water. Een succesvolle jacht noemden wij een 'vangst'. Ook agressieve interacties en prooiroof ('kleptoparasitisme') tussen Slechtvalken onderling of tussen valken en andere roofvogels of meeuwen, komen hier aan de orde.

RESULTATEN EN BESPREKING

Het aantal Slechtvalken dat Dekker op het wad bij Zwarte Haan zag was slechts drie in 34 uren van intensieve observatie, een gemiddelde van een per 11 uur. Ter vergelijking werd er een gezien per twee uur in het Noarderleech. De lage resultaten bij Zwarte Haan waren niet verwacht. Aanvankelijk meende Dekker dat deze locatie ideaal zou zijn voor het waarnemen van jagende valken, omdat de zeedijk hier direct grenst aan het open wad. Om zijn kansen verder te vergroten, koos hij bewust voor het tijdsbestek vlak voor of na hoogtij, wanneer de strandlopers en eenden het dichts bij de kust zaten. De relatieve schaarste aan valken suggereert dat zij het open wad vermijden. De verklaring ligt waarschijnlijk hierin dat Slechtvalken, die op wadvogels jagen, bij voorkeur gebruik maken van dekking om hun prooi ongemerkt zo dicht mogelijk te benaderen en te overrompelen. Alleen wanneer er onvoldoende dekking voorhanden is, gaan valken ertoe over om hun prooi ook op uitgestrekte, kale moddervlakten te bejagen (Dekker 1980).

Dit inzicht in de jachtgewoontes van de Slechtvalk werd verkregen na vele jaren veldonderzoek bij Beaverhills Lake, een prairiemoeras in Alberta, en in een Canadees kustgebied dat vergelijkbaar is met Noard-Fryslân Bûtendyks, met name Boundary Bay in British Columbia. Hier joegen de Slechtvalken zowel over het kilometerswijde wad als over de zee op Bonte Strandlopers. Daar hadden ze echter een vangstpercentage van 10–11%, terwijl ze vier keer zoveel succes (44%) boekten over de begroeiing van de kwelder (Dekker & Ydenberg 2004).

Voorgaande verklaart ook waarom er meer valken gezien werden in Noard-Fryslân Bûtendyks dan bij Zwarte Haan. De kweldervegetatie en zomerdijken gaven een jagende Slechtvalk volop gelegenheid gebruik te maken van dekking. In totaal telden we hier 128 valken in 260 uur. Vanwege het aanzienlijke verschil tussen beide gebieden zijn de gegevens van Zwarte Haan niet opgenomen in de tabel en figuren.

Het aantal observatie-uren per dag (2006 plus 2007) varieerde van 1 tot 14 (figuur 15.1) en het aantal observaties per dag van nul tot maximaal 1,41 per uur (figuur 15.2). In de eerste helft van de studieperiode was het aantal valken 0,56/uur vergeleken met 0,38/uur in de tweede helft van de periode. Deze af-name was waarschijnlijk het gevolg van doortrek. Een statistische analyse van alle gegevens laat een zwak dalende lijn zien (figuur 15.3).

Leeftijdsgroep en geslacht. In totaal, zagen wij 47 volwassen en 51 onvolwassen Slechtvalken, alsmede 30 van onbekende leeftijd (tabel 15.1). Voor zover ons bekend, maakt de bestaande literatuur uit Noord-Nederland zelden melding van overwinterende eerstejaars valken. Recente uitzonderingen zijn Van Dijk (1996), Teunissen (1997), Brandenburg & Riemersma (2002) en Marcus (2005). Afgaande op informatie van lokale wadvogeltellers, die in Noard-Fryslân Bûtendyks 's winters alleen volwassen valken zagen, hadden wij verwacht dat alle eerstejaars vogels eind november weg zouden zijn. Maar dat is niet gebleken. In de laatste zeven dagen van deze studie telden wij drie volwassen en vier onvolwassen Slechtvalken (tabel 15.1). Wel daalde het aantal eerstejaars valken in de tweede helft van de studieperiode, vergeleken met de eerste helft, van 0,37/uur naar 0,13/uur. Dit wijst erop dat er in de eerste 18 dagen meer doortrek van jonge valken plaatsvond dan daarna. Het aantal volwassen vogels zakte echter tezelfdertijd van 0,20/uur tot 0,14/uur.

Niet alleen in hun broedgebied, maar ook in overwinteringsgebieden zijn volwassen Slechtvalken sterk territoriaal, vooral de vrouwtjes en specifiek ten aanzien van hun eigen sekse (Dekker 1995, 1999). Vanaf bepaalde voorkeursplekken, die een goed uitzicht bieden, vliegen zij indringers van verre tegemoet, wat kan leiden tot een agressieve achtervolging. In Noard-Fryslân Bûtendyks hebben wij nooit luchtgevechten waargenomen tussen volwassen vrouwtjes onderling. Slechts eenmaal zagen wij er twee in zeer kort tijdsbestek. Zij zaten op enige honderden meters afstand van elkaar in Ferwerdaradeels Buitendijks. Bij onze nadering vloog de ene valk oostwaarts, de ander naar het westen. Naar vermoeden lag dit gebied op de grens van hun wederzijdse territoria.

In het Noarderleech en de Bokkepollen vonden wij geen aanwijzingen dat er meer dan een enkel volwassen vrouwtje voorkwam. Wel zagen wij drie maal dat de plaatselijke valk plotseling wegvloog met snelle vleugelslag om een andere valk te onderscheppen, maar dit betrof telkens een eerstejaars valk van ongeveer dezelfde afmetingen, dus ook vrouwtjes.

Canadese observaties hebben aangetoond dat vrouwtjes Slechtvalken de mannetjes niet verjagen en dat de twee geslachten soms dicht bij elkaar zitten, op enkele meters afstand. Dit werd ook opgetekend in Nederland (Van Dijk 1996; Klaassen 1996; Teunissen 1997; Wokke 2002; Marcus 2005). Het lijkt dan alsof er sprake is van een broedpaar dat samen overwintert (Beemster 1993), maar dat hoeft niet zo te zijn. Vrouwtjes tolereren het 'zwakkere' geslacht omdat ze daar voordeel van hebben. Mannetjes zijn kleiner, maar meer behendige jagers, en de vrouwtjes pakken hen zonder pardon de prooi af. Van onderlinge voedselconcurrentie is weinig sprake, althans in waterrijke gebieden, want de twee geslachten jagen over het algemeen op prooisoorten van verschillende grootte: het vrouwtje op eenden, het mannetje op steltlopers en zangvogels (Dekker 1980-2003; Van Dijk 1997).

In totaal telden wij in Noard-Fryslân Bûtendyks 26 keer een volwassen vrouwtje en 14 maal een volwassen mannetje. Onder de onvolwassen valken waren de mannetjes in de meerderheid (tabel 15.1). We zagen geen onderlinge vijandigheden of prooiroof tussen volwassen Slechtvalken, wel tussen jonge en oude dieren. Tweemaal streek een volwassen vrouwtje neer op een plek waar een eerstejaars valk opvloog. Dit gebeurde op te grote afstand om precies te kunnen zien wat er plaatsvond, maar waarschijnlijk liet de jonge valk een gevangen prooi achter. Vier maal zagen wij een volwassen mannetje dat net een prooi gevangen had, achtervolgd door een jonge valk. In minimaal een van deze gevallen, gaf de eerste zijn buit prijs, waarschijnlijk een Spreeuw, die in kwelderruigte viel en voor beide vogels verloren ging. Evenals volwassen vrouwtjes, zijn volwassen mannetjes territoriaal en wij zagen ze in Noard-Fryslân Bûtendyks hevig stoten op eerstejaars mannetjes.

Onvolwassen valken zijn onderling felle concurrenten. Afgezien van elkaar beroven, sluiten zij zich bij soortgenoten aan in de achtervolging van een prooi. In Canada joegen vaak drie of vier valken tegelijk op dezelfde strandloper. Tijdens de doortrek in Alberta en 's winters in British Columbia werd waargenomen dat zes Slechtvalken afwisselend op een vluchtige steltloper stootten totdat die uiteindelijk gepakt werd. Het waren dan altijd de vrouwtjes die met de buit gingen strijken (Dekker 1980, 2003).

Prooikeuze en jachtwijze. Wij zagen 33 aanvallen op potentiële prooien, waarvan er tien werden gevangen: zeven Bonte Strandlopers, een Smient, een Spreeuw, en een vogel van onbekende soort. Volwassen valken vingen zes prooien in negen jachtvluchten; onvolwassen valken slechts twee van de 15. Deze aantallen zijn te klein voor een veelbetekenende vergelijking met andere studies. Opmerkelijk is dat volwassen Slechtvalken een veel hoger succes (66%) boekten dan jonge valken (13%) en Slechtvalken van onbekende leeftijd (22%).

De meeste jachtvluchten begonnen als verrassingsaanvallen, laag over de kwelder of de waterkant. Om een aanstormende valk te ontwijken, gingen steltlopers of Spreeuwen op de wieken en probeerden hun belager af te schudden met abrupte wendingen. Op het laatste moment lieten zij zich in het water of de ruigte vallen. De Slechtvalk is echter zeer goed in staat een spartelende vogel uit het water te vissen. Eenden worden vaak gewoon op de grond gepakt, in kweldergras of in water dat niet diep genoeg is om onder te duiken (zie Dekker 1980, 2003).

Zes van de zeven Bonte Strandlopers werden gevangen na een korte achtervolging met 2–6 stootduiken. Eén werd er hoog in de lucht direct van achteren gegrepen. Driemaal zagen wij twee valken dezelfde strandloper achtervolgen, om beurten stotend. In alle drie gevallen betrof het twee mannetjes, de ene volwassen, de andere eerstejaars. Het doelwit van een bijzonder hardnekkige jachtpartij was een Kievit. Na ongeveer 27 vergeefse stoten, gaf de juveniele valk de aanval op. Een wendbare Kokmeeuw werd na vijf stootduiken alleen gelaten.

Bovenstaande lijst van prooien, waarvan de hoofdmoot uit Bonte Strandlopers bestaat, verschilt sterk van de plukresten die Oosterhuis & Kok (2003) op Griend aantroffen. Van de 56 prooien die zij vonden bestond iets minder dan

10% uit Bonte Strandlopers. Dit geringe aantal heeft er mogelijk mee te maken dat de overblijfselen van kleine vogels, zoals strandlopers, gemakkelijk gemist worden (zie Studiegebied en bespreking). Ter vergelijking: op Engelsmanplaat was de Bonte Strandloper met 19 exemplaren wel de meest voorkomende soort in 103 prooiresten (Smit 2000).

Prooiroof. Naast intraspecifieke prooiroof (van Slechtvalk op Slechtvalk) komt interspecifieke prooiroof (tussen soorten) algemeen voor in gebieden waar grotere roofvogels het terrein met de Slechtvalk delen. Op de westkust van Canada werden de talrijke zeearenden (Bald Eagles) gezien als reden waarom de valken zelden eenden vingen, want die zijn te zwaar om bij de nadering van een arend weggedragen te worden. In plaats daarvan joegen de kustvalken vrijwel uitsluitend op Bonte Strandlopers (Dekker 1998, 2003).

Tot onze verrassing bleek er een vergelijkbare situatie te heersen in Noard-Fryslân Bûtendyks. Hier betrof het echter geen arenden, maar Buizerds en Grote Mantelmeeuwen. Tegen deze agressieve rovers was een vrouwtjesvalk meestal wel opgewassen. Indien nodig, verdreef zij een Buizerd of meeuw met gekrijs en felle stoten, maar een mannetjesvalk liet zijn prooi meteen in de steek.

Terwijl mannetjes Slechtvalken zeer goed in staat zijn grote prooien te doden, zoals een Smient of een Pijlstaart, laten zij dat na als er rovers in de buurt zijn. In Canada werd waargenomen dat een onvolwassen mannetje twee talingen ving in five minuten en ze allebei verloor aan Buizerds. De valk zat toen voor lange tijd op een paaltje en werd pas na zonsondergang weer actief. Slechtvalken jagen vaak in de schemering, vooral op eenden. Eén van de redenen is dat eenden laat op de wieken gaan, op weg naar voedselgebieden. Een andere reden zou kunnen zijn dat valken, als het donker is, minder last hebben van rovers (Dekker 1980, 1987). Hustings & Van Winden (1998b) speculeren dat de recente neergang van het aantal overwinterende Slechtvalken in het Lauwersmeergebied een gevolg zou kunnen zijn van voedselconcurrentie en prooiroof van het groeiende bestand aan Buizerds en Haviken.

In Noard-Fryslân Bûtendyks ontdekten wij een mannetjesvalk die blijkbaar net een Wintertaling had geslagen die hij meteen prijs moest geven aan twee vechtende Buizerds. Een ander mannetje raakte een pasgevangen Bonte Strandloper kwijt aan een Grote Mantelmeeuw. Twee maal zagen wij mannetjesvalken die een prooi in nat gras of ondiep water gedreven hadden, maar hun pogingen de prooi te grijpen opgaven bij de nadering van mantelmeeuwen.

Mannetjesvalken die een strandloper vingen werden meteen benaderd door één of meer Grote Mantelmeeuwen. Wanneer de valk voldoende voorsprong had, liet hij de rovers al gauw achter zich en droeg zijn buit ver weg, vaak binnendijks. Landend in open terrein, zat hij dan minutenlang stil alvorens te plukken en indien nodig transporteerde hij zijn prooi opnieuw.

Het gewicht van een Bonte Strandloper is ongeveer 50 gram en een actieve valk eet 100–150 gram vlees per etmaal (Ratcliffe 1980). Dit betekent dat een valk elke dag minimaal twee of drie strandlopers moet vangen. Verliezen aan rovers voegen daar het nodige aan toe en vergroten de predatiedruk op de plaatselijke strandlopers.

Antipredatie-gedrag van Bonte Strandlopers. De kwelder Noard-Fryslân Bûtendyks was te breed om vanaf de zeedijk een goed gezicht op het wad te hebben. In de veronderstelling dat daar de beste kansen lagen om valken op strandlopers te zien jagen, liepen wij bij gunstig weer naar de buitenkant van de kwelder. De vaak harde wind en de afwezigheid van beschutting waren er echter debet aan dat we er zelden langer dan een uur bleven. De gelegenheid voor langdurige observatie werd echter aanmerkelijk beter als de wadvogels landinwaarts gedreven werden door springtij. Op 1 november 2006 en 9 november 2007 stond de gehele kwelder en de zomerpolder, na een uitzonderlijk hoog oplopende stormvloed, onder water. De Bonte Strandlopers foerageerden samen met vele honderden Kanoeten langs de voet van de zeedijk en op binnendijkse graspercelen. Nadat het waterpeil begon te zakken, werden de met slib bedekte grasvelden van het Noarderleech en de Bokkepollen gebruikt als hoogwatervluchtplaatsen. De vogels waren vanaf de zeedijk redelijk goed te zien.

Een bekende antipredatie-strategie van strandlopers is om bij de nadering van een Slechtvalk massaal op te vliegen en een dichte, bolvormige zwerm te vormen die woest heen en weer zwenkt, afwisselend zwart en wit opblinkend. De gangbare theorie is dat het individu veiligheid vindt in de zwerm en dat de grillige wendingen een roofvogel in verwarring brengen. In Noard-Fryslân Bûtendyks hebben wij valken nooit op dergelijke zwermen zien stoten. In plaats daarvan schenen ze zich te concentreren op de achtervolging van prooivogels die net van de grond of uit het water opvlogen.

Over-Ocean Flocking. In tegenstelling tot de Friese valken, belaagden Canadese 'kustvalken' geregeld dichte zwermen Bonte Strandlopers, niet alleen tijdens laag water over het kilometersbrede wad, maar ook tijdens hoogtij als de strandlopers, twee tot zes uur lang., boven de oceaan bleven rondvliegen. Dit bijzondere fenomeen staat bekend als *Over-Ocean Flocking* en wordt gezien als een antipredator-strategie (Dekker 1998). De strandlopers zouden de kuststrook en het binnenland vermijden vanwege het gevaar door valken en andere roofvogels overrompeld te worden. In plaats daarvan verkiezen ze het luchtruim boven de zee. Het gevolg is dat een hongerige valk genoodzaakt wordt de zwerm hoog in de lucht te achterhalen. Vervolgens gaat hij er dan herhaaldelijk loodrecht op stoten, totdat hij een van de strandlopers raakt of er een kan pakken die de zwerm verlaat. Gedetailleerde beschrijvingen van deze spectaculaire jachtwijze zijn te vinden in Dekker (1998, 2003).

Een vergelijkbaar hoogwatergedrag van overwinterende Bonte Strandlopers werd gerapporteerd in Jadebusen aan de Duitse Waddenzeekust door Hermann Hotker (2000). Hij nam dit vluchtgedrag waar op vijf van zijn 35 plaatselijke observatiedagen en noemde het *Airborne Roosting* of *Over-Sea Flocking*. Ook Hotker dacht dat het een antipredator-strategie was, hoewel hij zijn mening niet kon onderbouwen met gegevens over jagende roofvogels.

Eén van de vragen die wij ons stelden bij de aanvang van deze studie, was of er bij hoogtij ook Over-Ocean Flocking plaatsvond in Noard-Fryslân Bûtendyks. Dat bleek zeer zelden het geval. Ter verklaring en bij wijze van hypothese, geven wij de volgende vier redenen: (1) Op de meeste dagen rees het water niet voorbij de rijsdammen en bleven de Bonte Strandlopers in de ondieptes langs de kust; (2) Bij wassende vloed verhuisden zij naar de kwelder op plaatsen waar de begroeiing door begrazing zeer kort gehouden was; (3) Het gevaar door Slechtvalken overrompeld te worden (de predatiedruk) leek ons in Noard-Fryslân Bûtendyks minder sterk dan in Canada's Boundary Bay; En (4), een bijkomende factor zou gelegen kunnen zijn in klimaatsverschillen, vooral de in Fryslân overwegend zeer straffe wind. In tegenstelling tot Noard-Fryslân Bûtendyks is Canada's Boundary Bay beschermd tegen oceaanstormen door een gordel van bergachtige eilanden.

De onverwachte uitzondering vond plaats op 24 november 2007. De temperatuur was circa 8°C, er stond een zwakke aflandige wind en het was lichtbewolkt. Vanaf de zeedijk bij Hallumerhoek zag Dekker boven de zee zwermen strandlopers, die in traag veranderende formaties in beide richtingen parallel aan de kust trokken, een beeld dat sterk deed denken aan Boundary Bay. Om een beter zicht te hebben, reed hij naar de bunker. Het verschijnsel duurde ongeveer een half uur, tussen 8:45 en 9:30. Daarna streken de strandlopers geleidelijk neer langs de hoogwaterlijn.

Over-Ocean Flocking verschilt duidelijk van de welbekende heftige vluchtreacties van zwermen wadvogels die reageren op de nadering van een valk. In sterke tegenstelling daarmee vliegen Bonte Strandlopers tijdens over-ocean manoeuvres op een energiebesparende manier, met een vlinderachtige wiekslag afgewisseld door korte stukjes zeilen. Ze leunen als het ware als een vlieger op de wind. Het heeft veel weg van de trage vlucht van grote zwermen Goudplevieren en Kieviten, die hoog boven de polder blijven hangen, lang nadat de Slechtvalk, die hen aanvankelijk alarmeerde, uit het gezicht verdwenen is. Dit vlieggedrag van Goudplevieren en Kieviten, bekend bij alle Friese weidevogelaars, wordt algemeen gezien als een teken dat er onraad gesignaleerd is in de vorm van een roofvogel.

Het is de vraag of Over-Ocean Flocking ook in Noard-Fryslân Bûtendyks meer voor zal gaan komen als de predatiedruk van Slechtvalken en andere roofvogels (als Smelleken, Havik en Sperwer) verder zal toenemen. Een interessant toekomstbeeld schept ook de verwachte terugkeer van de Zeearend. Als die een algemene kustvogel gaat worden, zou dat, net als in Canada, van grote invloed kunnen zijn op de prooikeuze van de Slechtvalk.

Het geringe aantal jachtvluchten dat wij zagen, onderstreept onze mening dat Noard-Fryslân Bûtendyks een moeilijk terrein is voor dit soort onderzoek. De vele greppels, dobben en zomerdijken geven een valk volop de gelegenheid zijn lage jachtvlucht, niet alleen aan zijn prooi, maar ook aan het menselijk oog te onttrekken. De beste kansen om valken te zien biedt de buitenkant van de kwelder. Het probleem is echter dat er daar, in tegenstelling tot Boundary Bay, geen berijdbare weg parallel aan de kust loopt. Verder wordt het overzicht op het wad bemoeilijkt door de rijsdammen. Ook daar weet de valk zijn voordeel mee te doen door zijn prooi te overrompelen, maar voor de waarnemer leidt het vaak tot frustratie omdat een laagvliegende valk al snel aan het gezicht onttrokken wordt.

Conclusie. Wij kregen niet de indruk dat Noard-Fryslân Bûtendyks dicht bezet was met Slechtvalken. Deze conclusie is gebaseerd op het maximumaantal valken dat wij tegelijk zagen, vergeleken met gegevens uit Canada. Hierbij moet wel onderscheid gemaakt worden tussen de trektijd en de overwinteringsperiode. Tijdens de doortrek zijn valken niet territoriaal. Dan kunnen er vanaf een bepaald punt soms vier of vijf geteld worden, zittend op palen of op de grond, en binnen elkaars gezichtsveld (Beemster 1993). Dergelijke concentraties komen ook in Canada voor, niet alleen tijdens de voorjaarstrek in Alberta en British Columbia, maar ook 's winters in Boundary Bay. Daar achtervolgden soms vijf of zes valken dezelfde prooi. In Noard-Fryslân Bûtendyks zagen wij nooit meer dan twee valken tegelijk. Bovendien beperkten zij zich niet tot het buitendijkse gebied, maar vlogen ook over de zeedijk naar de landerijen.

Bovenstaande conclusie wordt versterkt door het relatief gering aantal keren (zeven) dat wij Slechtvalken Bonte Strandlopers zagen vangen: slechts een per 37 uren observatie. Ter vergelijking: in Boundary Bay werd een vangst per tien uur geregistreerd wat opliep tot vier vangsten per tien uur vlak na hoogtij, wanneer de hongerige strandlopers dicht bij de kwelder neerstreken om voedsel te zoeken (Dekker & Ydenberg 2004).

Nawoord. Tussen 8 december 2007 en 3 januari 2008, heeft Albert Ferwerda het onderzoek in Noard-Fryslân Bûtendyks voortgezet met 22 uren van observatie verdeeld over acht dagen. In totaal zag hij vijf Slechtvalken: drie volwassen, een onvolwassen en een vogel van onbekende leeftijd. De laatste Slechtvalk, een eerstejaars vrouwtje, werd gesignaleerd op 20 december. Deze lage aantallen doen vermoeden dat de Slechtvalken die wij tussen 30 oktober en 4 december zagen, voor het overgrote deel doortrekkers waren. Ook het aantal steltlopers en eenden was eind november sterk verminderd, althans in de zomerpolder en kwelder, vooral na enkele dagen van lichte vorst medio december.

Tabel 15.1. Aantal waarnemingen van volwassen en onvolwassen Slechtvalken en valken van onbekende leeftijdsgroep in NFB tussen 30 oktober en 4 december, 2006 en 2007.

Volwassen vrouwtjes Slechtvalken	26
Volwassen mannetjes	14
Volwassen valken van onbekend geslacht	7
Aantal volwassen Slechtvalken	47
Onvolwassen vrouwtjes Slechtvalken	13
Onvolwassen mannetjes	21
Onvolwassen valken van onbekend geslacht	17
Aantal onvolwassen Slechtvalken	51
Slechtvalken van onbekende leeftijd en geslacht	30
Totaal 128	

Figuur 15.1. Aantal uren van observatie per dag in NFB. Gegevens voor dezelfde datums zijn samengevoegd voor 2006 en 2007.

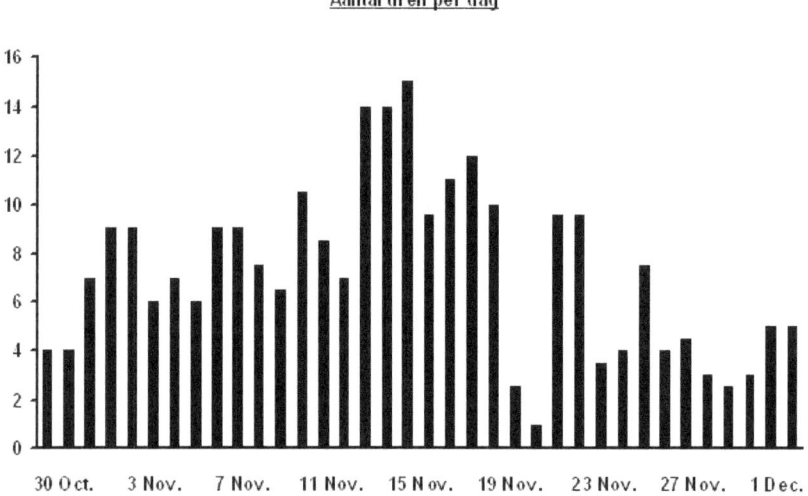

Figuur 15.2. Aantal Slechtvalken per uur van observatie in NFB. Gegevens voor dezelfde datums in 2006 en 2007 zijn samengevoegd.

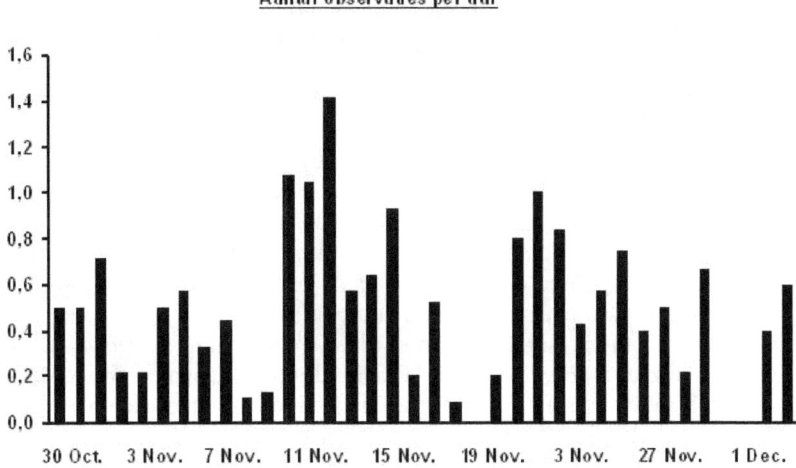

Figuur 15.3. Statistische daling van het aantal Slechtvalken per uur in NFB. Gegevens voor dezelfde datums in 2006 en 2007 zijn samengevoegd.

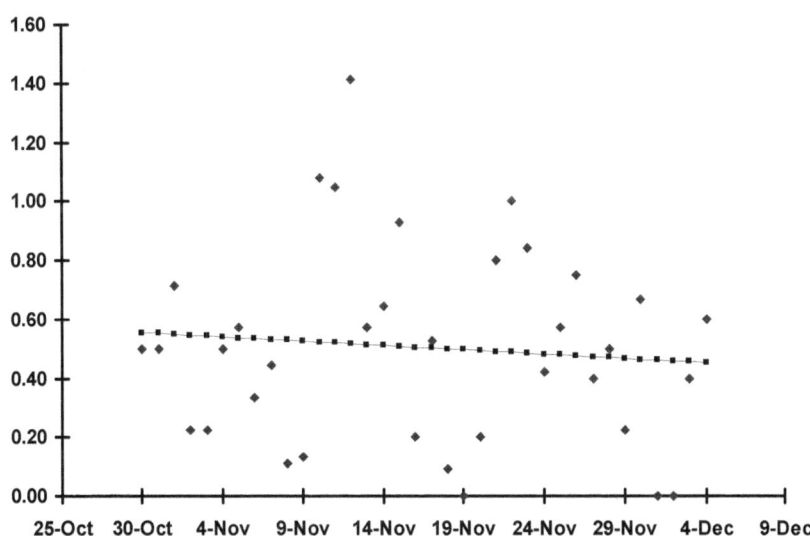

Abstract. Migrations and foraging habits of Peregrines migrating and/or wintering on the Frisian coast of the Wadden Sea in the Netherlands, with special reference to high-tide roosting behaviour of wintering Dunlins (*Calidris alpina*).

During 59 days between late October and early December, 2006 and 2007, we observed Peregrine Falcons migrating and wintering in a 4.100 ha strip of coastal wetland lying outside the sea wall near Hallum, Friesland. Of 128 sightings, 40% were first-year immatures, 37% adults, and 23% of unidentified age. The proportion of females to males was roughly one to one in the total sample, two to one in adults, and two to three in immatures. Total sightings dropped from 0.56/hour in the first half of the study period to 0.38 in the second half, suggesting that most falcons were passing through on their southbound migration. Supplemental observations on eight days between 8 December 2007 and 3 January 2008 produced another five sightings, of which three were first-year immatures. Thirty-three attacks on potential prey resulted in ten captures: seven Dunlins, one wigeon, one starling, and one unidentified bird. Peregrines with prey were klepto-parasitized by Great Black-backed Gulls (*Larus marinus*) and Common Buzzards (*Buteo buteo*). At high tides that inundated all intertidal mudflats, the Dunlins roosted on short-grazed meadows on either side of the sea wall. Over-ocean flocking by Dunlins was seen only once. The phenomenon ended after 0.5 hour, but might have begun before it was noticed.

Footnote. This paper includes the observation that Golden Plovers and Lapwings engage in an antipredator strategy that is reminiscent of over-ocean flocking. When alarmed by a raptor, the plovers rise high into the sky and keep flying about in flocks until they descend again after the danger has passed. The phenomenon is familiar to Dutch birdwatchers, and no doubt to British birders. Ratcliffe (1980:157) includes a quote from Ireland to the effect that hunting Peregrines induce a state of near-panic in some birds, and that "the Golden Plover and Lapwings rise and keep flying about at an immense height, sometimes for hours." I suggest that this flight behaviour is worthy of further study. Perhaps a suitable term for it might be "high-sky flocking."

CHAPTER 16

Discussion and Synthesis

Its near extirpation half a century ago and subsequent spectacular recovery have made the Peregrine one of the most celebrated birds of all time, in large part due to its exposure in the news media and on the electronic worldwide web (Kiff et al. 2008). Coupled with this increase in interest, among the scientific community as well as the general public, there came a need for more knowledge of the wild falcon, which motivated me to conduct detailed field investigations that began in 1960, and eventually included nine widely different study areas at various times of the year. The major findings of this research, conducted intermittently over 48 years, are reviewed here and placed in a contemporary perspective.

In the following sections I summarize what I learned about: (1) falcon hunting success rates; (2) their hunting methods and individual variations; (3) the evasive techniques of the major prey species; (4) intra- and interspecific kleptoparasitic aggression; (5) parent-fledgling interactions of Peregrines; and (5) their migratory behaviour. In addition, I make interspecific comparisons in hunting methods between four species of falcons, and I discuss the non-lethal effects of falcon hunting activity in respect to the ecology of fear.

Hunting success rates. My 1980 sample of 674 hunts and 52 kills, obtained at Beaverhills Lake in Alberta, was the largest set of first-hand data on foraging flights by Peregrines available at that time (Chapter 3). Its success rate of 7.7% was nearly equivalent to the 7.5% in 253 attacks observed in Sweden by Rudebeck (1950), whose report had been severely criticized by raptor authorities as being unrealistically low for a falcon that was considered a superior avian hunter (Fischer 1967; Brown and Amadon 1968; Ratcliffe 1980). Their objections were that Rudebeck might have confused genuine hunting attempts with playful interactions and mobbing of other birds, such as large raptors, that should not have been considered as suitable Peregrine prey. Ratcliffe (1962) questioned the intent of some of the unsuccessful hunts that he observed and suggested that the falcons were only playing or practising.

By contrast, Treleaven (1980, 1998) recorded a rate of about 60% in 55 hunts by breeding adult Peregrines along the coast of Cornwall. He argued that hunting takes place at two levels of intensity, and that low intensity hunts should not be considered as serious attacks. In my view, the explanation for the great difference in the above values is simply that Rudebeck's and my data pertain to migrating Peregrines, which include first-year immatures that cannot be expected to have comparable hunting skills or a similar level of intent as breeding adults that need to forage for a mate and nestlings (Dekker 1980).

The Beaverhills Lake study was first to give separate hunting success rates for various age classes of falcons. Although the sample of hunts by fall immatures migrating through central Alberta was small, their hunting success of 2.4% was indeed lower than the 9.8% of adults. A significant disparity between young and old (26.8% vs. 9.0%) was also evident in the much larger sample of hunts and kills collected during fall and winter at Boundary Bay on the coast of British Columbia (Chapters 7–9). However, it was interesting to note that first-year Peregrines returning from their wintering grounds had evidently gained enough experience to approach adult success, as observed at Beaverhills Lake. With respective rates of 7.1% and 9.8%, the difference between them was not statistically significant (Chapter 3).

To us humans, it stands to reason that raptors gain in expertise as they become older. However, to my knowledge, the only data in support of that notion and pertaining to wild Peregrines are my own, collected at Wabamun Lake in central Alberta (Chapter 10). Between its first and seventh nesting season, the pair's hunting success rose from 21.9% to 39.1%. And by the tenth year, when the female was 12 years old, her overall success rate had climbed to just over 46%. These falcons hunted Franklin's Gulls (*Larix pipixcan*) that flew by their nest box attached to the chimney of an electricity generating station. From its high perches, or soaring above the plant, the pair stooped selectively at immature gulls, which made up 85% of gulls taken.

Quite similar to the hunting habits of the Wabamun Peregrines, Treleaven's falcons also launched very successful attacks from high vantage points, but their main prey was Rock Doves flying over the sea below the falcons' nesting cliffs. The vulnerability of land birds over water was already pointed out by Herbert and Herbert (1965) and emphasized by Hunt et al. (1975).

However, there appears to be no hard and fast rule governing Peregrine hunting success, which may vary from one study to the next even if they involve the same kind of prey. For instance, at another location on the seacoast of Britain, Parker (1979) reported a success rate of 16% in 113 attacks on Rock Doves. In California, adult Peregrines scored 40% in 1485 attacks on Rock Doves (Palleroni et al. 2005). Unfortunately, the details of these hunts and the methods used are incompletely described. By contrast, Jenkins (2000) recorded only 9–12% success of African Peregrines, which launched 1318 hunts from their nesting cliffs at doves and passerines. The success of these falcons was correlated with cliff height.

Evidently, as noted by Palmer (1988), the capture rate of the Peregrine varies depending on many factors. These include: (1) habitat and season; (2) the kind of prey hunted; (3) the age and expertise of the falcon; and (4) its motivation. In a review of 23 different studies, White et al. (2002) found an average rate of 23.7%. The highest success ever reported – about 95% in 170 hunting flights – was achieved by a resident adult male Peregrine in New Jersey. Apparently, this exceptionally capable falcon focussed on vulnerable targets such as jays migrating over coastal marshes (Cade 1982).

Against the above background, it is most interesting that the lowest recorded success rate, even lower than the 7.5–7.7% reported for migrating Peregrines by Rudebeck (1950) and me, was obtained at the Tyninghame estuary in southeast Scotland. There, only 6.7% of 368 attacks by wintering falcons were successful (Cresswell and Whitfield 2000). The important factor here is that these falcons hunted shorebirds in their own preferred environment instead of taking advantage of individuals that happen to be outside their typical habitat. It is in these natural wetland habitats, where waterbirds and Peregrines are in a sense on even terms, that we can learn to understand the dynamic interactions between raptors and their prey.

By way of comparison to Cresswell's data, my long-term studies of wintering Peregrines on the Pacific coast of British Columbia produced an overall success rate of 14% in 652 attacks on Dunlins (*Calidris alpina*). Over the open mudflats or ocean, the rate dropped to 10–11%. However, if these Peregrines concentrated their attacks on Dunlins feeding or resting in or close to the shore, the predation rate shot up to 44% (Chapter 7–9). This is in the same order of magnitude as some of the highest rates reported above, and again illustrates that Peregrines are most successful if they attack prey in vulnerable positions away from cover.

The success rate of Peregrines attacking shorebirds at Beaverhills Lake was 9.0%, based on 400 hunts (Chapter 3). During the 15 years of that field study, I also saw 153 duck hunts, which had a success rate of 8.5%. However, after eight additional seasons of watching, my sample of duck hunts had increased to 275 and their success rate had risen to 9.0%, the same as the rate on shorebirds.

A falcon's intent of making a quick kill can be expected to vary with the state of its appetite. It is fair to say that a hungry falcon is a deadly falcon. Serious intent can also derive from the need to forage for a mate or brood. My 1980 paper was first to advance the idea that breeding Peregrines are more successful that non-breeding falcons, an idea that has since been supported by a much larger range of data including several studies of falcons wintering in Sweden (Roalkvam 1985).

To sum up, the wide spread in values across the publications quoted above indicate that the success rate of the Peregrine is indeed dependent upon many factors, of which motivation is of critical importance. The fact that migrating or wintering falcons appear to have low capture rates, compared to nesting falcons, should favour the prey species, because they have a nine to one chance of surviving an attack providing they employ effective anti-predator and escape tactics.

The section below describes the various methods Peregrines use in hunting their principal prey species and how these prey manage to dodge their attackers.

Hunting methods. My long-term observations differ in several major aspects from established theories about the Peregrine's hunting habits. For instance, I found very little support for the frequently mentioned claim that the Peregrine pursues its target openly, high in the sky, and smacks it a hard blow that sends it to earth mortally wounded or dead (Brown and Amadon 1968; Ratcliffe 1980; Godfrey 1986). Reports to that effect appear to be largely based on falconry birds that are trained to wait on high until the falconer and his dogs flush a grouse or partridge from its hiding spot in grass or heather. Coming down at great speed, the falcon quickly overhauls its target and rakes it with its protruding hind claws. In the wild, similar aerial strikes do happen, but very rarely, because the falcon's primary goal is to secure its food. Allowing the prey to fall down comes at the risk of losing it in dense vegetation. Furthermore, birds such as partridges do not normally rise under a falcon, but stay crouched until the danger has passed. By way of another example, Canadian falconers often fly their Peregrines at ducks that are scared up from ponds using dogs. Under natural conditions, a duck would never flush from water if there were a falcon in the air above. This makes the circumstances under which a falconry bird hunts very different from those of wild falcons. In my experience, in nearly all of the aerial captures observed in detail, the Peregrines just seized the prey in their feet and brought it down to a suitable plucking place or carried it to their nest.

Surprise attacks. Like all mammalian and avian predators that hunt prey closely matching them in speed and manoeuvrability, the Peregrine's first objective – before it can attempt to seize its prey – is to get as close as possible to the target before it flees. The reason is simple. Birds that have had a timely warning of the falcon's approach usually escape capture by dodging astutely or by taking cover in water or dense vegetation in which they are safe from further pursuit. Therefore, the attacking falcon's priority is to stay out of sight until the very last moment. This makes stealth attacks the most widely used tactic for bird-hunting raptors in North America and Europe (Page and Whitacre 1975; Dekker 1980, 1988, 1995, 1998; Cresswell 1996).

In some form or other, Peregrines everywhere try to capitalize on the element of surprise, particularly in habitats where the vegetation allows a low flying falcon to hide behind a screen of tall grasses or reeds. At Beaverhills Lake stealth attacks made up 70% of all hunts (Chapter 3). Approaching at speed over the open fields and pastures surrounding the lake, the falcons achieved surprise by: (1) sustained low flight; (2) long-range descent after the target had been spotted well ahead; or (3) a stoop with wings flexed from high soaring flight. In all three methods, the final approach took place very low over ground or water.

The stoop from soaring flight could be oblique or nearly perpendicular but the terminal stage of the attack was characterized by an abrupt change of direction. This method had a success rate of 11.0%. The long distance descent was rated at 8.0%, and the low race, in which the feeding or resting prey is encountered by chance, had the lowest success rate of 4.7%. In this latter method the prey could literally be *taken by surprise,* which is a commonly used phrase that originates from falconry. For detailed descriptions of various kills made by way of these surprise methods, see Chapter 3.

Like the migrating Peregrines at Beaverhills Lake, falcons wintering on the coast of British Columbia aimed nearly all of their attacks on prey sitting on the ground or in shallow water, and they used similar stealth tactics, except their opportunities for effective surprise were low due to the open character of the mudflats and the ocean (Chapters 7–9. In this coastal habitat, only 35% of 652 hunts aimed at Dunlins began as stealth attacks. In at least 60% of all other hunts the falcons were forced to approach their prey openly, because the Dunlin flocks had spotted the falcon in time and flushed from the ground. Surprise hunts on the coast had a success rate of 23.6%, significantly higher than the 9.1% of open attacks on Dunlins flying over the mudflats or ocean. Interestingly, the Peregrine's success rate shot up to 44% if it used the sea dike or marsh vegetation as cover to attack Dunlins roosting on the shore. The above figures illustrate the importance for a hunting falcon of staying out of sight as long as possible.

Stealth methods were also the principal tactic of Peregrines that wintered in an agricultural region of Vancouver Island, British Columbia. There, they hunted mainly ducks including American Wigeon (*Anas americanus*) and Northern Pintails (*Anas acuta*) that spent most of their day on flooded fields. To feed on grass the ducks had to walk farther and farther away from the water. Watching from its usual perch in a high tree, the falcons waited for the right moment to launch a descending surprise attack. However, in the majority of cases, the ducks rushed back to the safety of the pool before the Peregrine had come close enough to grab one of them (Chapter 5).

A most interesting variant of the stealth method was used by falcons resident on Langara Island, situated well off the northwest coast of British Columbia just south of Alaska. Here, in this pristine wilderness environment, nesting pairs of Peregrines employed a unique way of capturing small sea divers, such as auklets and murrelets that seldom take to the air. Approaching fast and very low over the water, covering the last ten or more metres in sail flight, the falcon struck the swimming bird in passing a glancing blow, raking it with its claws. If the prey had not dived in the nick of time and was indeed hit, the falcon quickly turned back and attempted to grab the crippled alcid, lifting it out of the water. These coastal falcons also made use of waves and partly submerged rocks to conceal their stealth approach (Chapter 6).

A preference for taking their prey by surprise was also common of the Peregrines hunting Rock Doves flying along the British coast. Starting from their high nesting cliffs, these falcons tended to overhaul the target bird directly from behind in its blind spot (Treleaven 1980, 1998). Surprise was even a component of the high altitude hunts of the Peregrines that bred for ten years on an industrial plant at Wabamun Lake, Alberta. Working in tandem, the pair soared up to altitudes estimated to be well over 1 km. Upon spotting a distant gull, the female made a fast attack on a descending trajectory, ending up roughly at the target level, while the male stayed high. If his mate missed the dodging prey, the male stooped down perpendicularly to seize the gull directly from above (Chapter 10).

Quite apart from the surprise tactics described above, individual Peregrines are quite versatile in the way they hunt or choice of prey (Fischer 1967; Cade 1982).

In British Columbia, I watched adult males climb high into the sky to meet flocks of teal, and force them down to the ground (Chapter 5).

The most persistent in aerial pursuit are typically the juveniles during their first fall. As if to compensate for their lack of skill in surprise attacks, they chase shorebirds over distances that may exceed several kilometres and rise to great heights, often disappearing from view, which makes it difficult to determine the outcome and success rate of these long hunts. During the chase, the falcon makes repeated passes and swoops at the dodging target, until it is let go or plunges steeply to earth. In many cases, the hard-pressed shorebird eventually finds cover in reeds or other vegetation, which makes these high-energy pursuits quite unproductive. Shorebirds commonly drop into water to evade an attacking falcon. These are quickly plucked from the surface by adult Peregrines, but fall immatures have great difficulty. Nevertheless, if necessary they will swoop at the splashing prey 10–35 times until it is secured (Chapters 3, 7, 9).

By contrast, I saw three adult Peregrines pounce on small rodents, and two others went down on carrion (Chapter 3). Along the Red Deer River, breeding Peregrines fed ground squirrels to their nest young (Chapter 12). Peregrines also inspect and clean up prey remains left by other falcons. In addition, all large raptors, including the falcons, are quick to take advantage of every opportunity to rob smaller raptors of their catch (Chapter 3, 12).

Klepto-parasitism. The common occurrence of prey theft between raptors, both intra- or interspecific, has received scant attention in the literature (White et al. 2002). The exception is Cresswell and Whitfield (1994). During all of my field studies, I have been impressed by the reality of the bully's motto that "might is right." Klepto-parasites force the Peregrine to kill more than its own food requirements. In British Columbia and in Alberta, I saw falcons kill two ducks in succession and lose them at once to other raptors (Chapters 3, 4, and 5). On Vancouver Island, at Langara, and at Boundary Bay, prey-carrying Peregrines were immediately chased by eagles, and sometimes by Gyrfalcons or Prairie Falcons. At Beaverhills Lake and in the Netherlands, male Peregrines were robbed of prey by buteo hawks and large gulls (Chapter 3, 5, 6, and 16). By the same rule, the Peregrine hijacks all raptors smaller than itself, including harriers, small accipiters, and Merlins. Intraspecific klepto-parasitism is also common. Both at Boundary Bay and Beaverhills Lake, female Peregrines routinely robbed males. To avoid aggressive females and other pirates, male Peregrines often take small prey high into the sky and feed while soaring (Chapters 3 and 10).

Prey species. The prominence of surprise in the hunting methods of a raptor that is considered the fastest creature on earth is all the more remarkable in view of the fact that the Peregrine's level flying speed is greater than all birds it preys upon, with the possible exception of some champion racing pigeons (Treleaven 1980). By making use of gravity, the stooping Peregrine can close in with any bird it wants, however the interesting fact is that nine out of ten prey attacked manage to make their escape as long as they are in their preferred habitat and take the correct precautions. The cumulative success rate of 11.7% reported in this study (Table 16.1) reveals that eight or nine out of every ten prey attacked managed to evade the falcon. How the various prey species were hunted is summed up in the following sections.

Pigeons. Whyte et al (2002:8) wrote that "in temperate continental latitudes Columbidae (pigeons and doves) may be most frequently taken and perhaps most important by biomass." In North America and elsewhere in the Peregrine's worldwide range, domestic and feral Rock Doves are indeed the main stay for urban falcons (Ratcliffe 1980, 1993; Frank 1994). However, at Beaverhills Lake, where I observed migrants from northern populations, none of the nearly one thousand hunts were directed at doves, and in all of my 44 years there I have found only a single prey remains of a homing pigeon at the lake. Although a few passerines are taken, the main prey of wild falcons migrating through central Alberta and wintering on the west coast of Canada are waterbirds, with waders and ducks in first and second place, and gulls in third (Table 16.2).

Ducks were most vulnerable when flying overland and could be seized in the air or on the ground (Chapters 3 and 4). When attacked over water, they splashed down to safety. Teal and other small waterfowl were light enough to be carried along by the falcon and could be seized over water as well as on the ground. Swimming ducks dived if attacked. Nevertheless, if the pond or lake was not deep enough for a duck to completely submerge, it might be seized in the shallows and dragged to dry ground, or fed upon while the carcass was lying in water. Examples of typical duck captures are described in Chapters 3 and 4.

The local presence or absence of Peregrines is reflected in duck behaviour. After a single pass by a Peregrine, open stretches of lakeshore where ducks haul out to loaf become deserted. Instead, the ducks move away from the shallows and spend the day on deep water. The timing of their foraging flights may also be altered by the threat of predation. During the spring migration at Beaverhills Lake, when Peregrines are common, ducks flew to and from inland nesting and feeding sites at nightfall. Similarly, at Boundary Bay, wintering wigeon departed on their daily foraging flights after sundown (Dekker 1999). Pintails and wigeon wintering on Vancouver Island were very much affected by the danger posed by Peregrine attack and restricted their feeding bouts to the safest sites closest to water (Chapter 5). By contrast, in the Netherlands, during the years when Peregrines were rare, wigeon flocks could be seen on grassland at any hour of the day. However, now that the Peregrine is a common winter resident on the Dutch coast, wigeon have become wary and cautious, rushing back to deep water at the least sign of danger as I noticed during my 2007–2008 field studies in Friesland (Chapter 15). These observations support the theory that the recent return of the Peregrine has been a major determinant in changing the behaviour of their prey species, including ducks.

Shorebirds use two diametrically opposed tactics to cope with danger: fleeing or hiding. Species such as the Pectoral Sandpiper (*Calidris melanotos*) and the Least Sandpiper (*Calidris minutilla*), which tend to frequent muddy ground or short grass, crouch and freeze at the approach of raptors (Burns and Ydenberg 2002). The streaked dorsal feather pattern of these sandpipers provides camouflage on a dark background. By contrast, the Semipalmated Sandpiper (*Calidris pusilla*), which is the most common migrant at Beaverhills Lake, has a plain dorsal colour and feeds in the shallows. It does not crouch when the alarm is sounded. Instead, it takes to the sky and joins its fellows in dense defensive formations. So does the Dunlin, the most numerous sandpiper wintering on the Pacific coast (Chapters 7–9). It rarely crouches and instead takes wing to outmanoeuvre its attackers. By rising into the sky and forming dense flocks, these birds actually draw attention to themselves. Yet, based on the large data set obtained in this study it is clear that the majority of sandpipers manages to evade even the most determined of Peregrines. The falcon's best chance lies in taking the prey by stealth. Surprise attacks were the most successful method on the coast and at Beaverhills Lake (Chapters 3 and 9).

How individual sandpipers can influence their own vulnerability and chances of escape has been a subject of recent research. For instance, high body mass in prey increases predation risk and there is a critical trade-off between food and safety (Gosler et al. 1995; Dietz et al. 2007). At Beaverhills Lake, I noted that Peregrines commonly caught Pectoral Sandpipers that failed to use escape tactics normal for their kind (Chapter 3). Although they flushed in time at the falcon's fast approach, these sandpipers dropped back into the grass at the last moment and stayed down, instead of taking off at once again into the opposite direction as soon as the Peregrine overshot the spot. In retrospect, I realize that this lethal mistake might well fit the trade-off theory that abundant food supplies may come at increased predation risk (Chapter 5). These Pectoral Sandpipers had interrupted their northward migration to take advantage of a massive hatch of what is locally known as lake flies: a large (10–15 mm) species of chironomid midge (Dekker 1991, 1998). Early on during my observations, I noted that lone sandpipers were captured more often than individuals in flocks (Chapter 3), which supports the experimental findings of Powell (1974) that there are social as well as safety benefits in flocking together. Cresswell (1994) presented field evidence that flocking is an effective anti-predation strategy for wintering waders in Scotland.

Over-ocean flocking. After three decades of watching the interaction of shorebirds and falcons at Beaverhills Lake, I was well prepared to begin observations at Boundary Bay, a section of the Fraser River Delta on the Pacific coast of British Columbia. There, I collected a very large sample of hunts and kills, and on the first day I recorded the phenomenon of over-ocean flocking (Chapters 7, 8, and 9). As reported in this study, during high tides that inundated all intertidal habitats, the vast majority of the locally wintering Dunlins (*Calidris alpina*) flew out over the ocean and remained airborne for 1.5–6.5 hours until the tide receded again. They then landed on the emerging mudflats and soon began to forage. I interpreted this flight phenomenon as an anti-predation strategy. The explanation was that the Dunlins were unwilling to spend the high tide period on roosting sites in the vegetated shore zone. There, the danger of surprise attacks by falcons proved to be greatest. This hypothesis was strongly supported by the comparatively very high success rates of Peregrines along the edge of the salt-marsh as compared to hunts over the extensive open mudflats and the ocean. In addition to the above increase in hunting success rate, I also recorded four times as many Dunlin captures per hour of observation just after high tide as before high tide or at ebb time. Apparently, after returning from their over-ocean flocking flights, the sandpipers were at their most vulnerable, and their need to eat had shifted the trade-off balance between food and safety towards higher risk (Chapter 8).

In Dekker (1998), I speculated that over-ocean flocking must have been an innate and routine behaviour in the past prior to European arrival in the region; because at that time the Fraser River Delta was likely overgrown with vegetation, and Peregrines can be expected to have been quite common (Chapter 7). However, a drastic habitat change was initiated about a century ago, after settlers began building dikes, digging drainage ditches, and installing pumping stations. The newly created open fields and pastures allowed the Dunlins, during the years of Peregrine decline, to fly inland at high tide as well as to roost along the edge of the salt marsh. In further support of this hypothesis, during the pesticide era, over-ocean flocking was also unknown on the Dutch coast of the Wadden Sea. There, too, the explanation is related to human-wrought changes of habitat. In winter, the coastal polders of Friesland include vast expanses of ploughed fields and meadows closely cropped by livestock, which provide shorebirds with suitable roosting sites. Over-ocean flocking is still rare in the Netherlands, even though Peregrines have made a comeback. However, they are not yet as numerous as at Canada's Boundary Bay (Chapter 15).

An additional reason why over-ocean flocking is common at Boundary Bay is related to the ubiquitous presence of Bald Eagles, which discourages locally wintering Peregrines from preying on ducks. To avoid klepto-parasitic eagles, the falcons stopped hunting waterfowl and instead concentrated their foraging on much lighter prey species. This hypothesis was supported by a comparison of prey captured in fall and mid winter. In October and November, when Bald Eagles were practically absent from the bay, I commonly saw female Peregrines capture ducks. However, in December and January, after the arrival of 100–150 wintering eagles, Peregrines had stopped hunting waterfowl and switched to Dunlins, which were light enough to be carried away out of reach of most pursuers (Chapter 9). The resulting increase in predation pressure on the Dunlins probably added extra urgency to their departure on over-ocean flocking. At localities where the phenomenon is rarely seen, eagles are absent and falcons catch a wide variety of prey (Chapter 15).

The ecology of fear. As recently reviewed by Cresswell (2008) the non-lethal effects of predation may well be as important or perhaps even more important than the actual mortality caused by predators. This applies to birds as well as mammals. Lethal predation by large carnivores is considered a major determinant in the demographics of ungulate populations (Mech and Peterson 2003). However, Creel et al (2007) postulated that non-lethal effects might be even more costly in terms of prey fitness and reproduction. In Montana, wapiti (*Cervus elaphus*) respond to wolves (*Canis lupus*) on a spatial scale of several kilometres. In Canada's Jasper National Park, wapiti and wolf presence were inversely correlated (Dekker 2008). Over a two month period in 2000, the local herd fled eight times back and forth across a major river and a busy highway, a distance of >2 km. Wolves also had an obvious impact on habitat choice by Rocky Mountain bighorn sheep (*Ovis canadensis*), restricting their winter range to one open slope. There, attacking wolves could be timely spotted, but the vegetation on this safe site became heavily grazed by late winter (Dekker 2002, 2008). These dynamics parallel the findings of ornithologists who have investigated the predation-related dynamics of avian communities.

Foraging under the threat of predation has become a major subject for field studies and experimental research (Stephens et al. 2007). Under the title "Foraging and the Ecology of Fear" Brown and Kotler (2007) have reviewed a wide variety of studies on the indirect cost of predation on songbirds. Studies in British Columbia revealed that the threat of Peregrine attack had caused shorebirds to alter their migration routes and avoid high-risk feeding or roosting sites (Ydenberg et al. 2004). Mediated by fear, the trade-off in all of this is that safety comes at the cost of a decrease in food resources. To reduce predation danger, foraging mammals as well as birds tend to under-utilize dangerous areas and over-use safe areas (Prins and Iason 1989; Cresswell 2008).

The data on hunting tactics of Peregrines, reported in this thesis, lent support to the hypothesis posed by Lank et al. (2003) that the return of the Peregrine has caused sandpipers migrating in British Columbia to change their choice of feeding sites. They were now spending less time in the most dangerous spots close to vegetation, and more time in wide-open habitats where an attacking predator could readily be spotted, although nutrient levels might be lower there than close to the saltmarsh shore (Pomeroy 2006; Pomeroy et al 2006).

Similarly, over-ocean flocking can be seen as a danger management behaviour providing relative safety from raptor attack but coming at the cost of increased energy expenditures. Over-ocean flocking was rarely reported at Boundary Bay during the 1970s and 1980s (Ydenberg et al. 2009). Then, these flights might not have been necessary because falcons were scarce or absent, and the Dunlins were known to spend the high tide interval by roosting on the edge of the saltmarsh or on inland fields. In recent winters, some inland feeding and roosting flights still take place on rainy and windy days (Chapter 7) and during the night (Evans-Ogden et al. 2005).

Interspecific comparisons. How the hunting methods and success rates compare between different species of falcon is an interesting area of raptor ecology about which there is very little mention in the literature. During my long-term field investigations, I have observed Merlins, Prairie Falcons, and Gyrfalcons in wetlands and agricultural plains, as well as in the Alberta foothills and alpine tundra of the Rocky Mountains. Everywhere, the principal hunting method of the above three species of falcon was to approach prey by stealth, much like the Peregrine. The surprise tactics of the Merlin and the Gyrfalcon are well known (Page and Whitacre 1975; Palmer 1988; Clum and Cade 1994; Cresswell 1996; Cade et al. 1998). In her comprehensive review of the food habits of the Prairie Falcon, Steenhof (1998) states that it flies low when hunting open-country passerines. However, she did not mention that the common habit of this falcon is to soar up to a great height during suitable weather conditions and launch long-range stoops terminating in a low surprise attack on ground-based prey that includes rodents as well as birds. This strategy is detailed in Dekker (1982) and parallels the Peregrine's style. In this thesis I present additional new information on Prairie Falcons preying on feral pigeons in a city environment, and I compare their methods to Gyrfalcons hunting pigeons at the same locality (Chapter 13).

The foraging habits of the Peregrine and the Merlin have been compared by Rudebeck (1959), Buchanan (1996), and Cresswell (1996). Both of these falcons are aerial hunters of birds, but the Peregrine is capable of capturing prey of a size that is beyond the Merlin ((Figure 16.2). My long-term observations at Beaverhills Lake provided an opportunity to compare their respective success rates on prey species in their overlapping size range (Chapter 11). In hunting sandpipers, the capture rate of both falcons proved to be quite similar, but the Merlin was significantly more successful than the Peregrine in catching small passerines. This fits the theory that a falcon pursuing prey in the air should have a body mass close to that of the prey in order to match as closely as possible the speed and manoeuvrability of that prey (Anderson and Norberg 1980; Cade 1982).

In a parallel equation, the success rate of the Prairie Falcon in hunting city pigeons was significantly greater than that of its larger congener the Gyrfalcon. Furthermore, they showed very different methods, with the Prairie Falcon stooping from above at individual pigeons in the flock, whereas the Gyrfalcon utilized the momentum of a downward pass to swoop back up into the compact mass of birds, apparently grabbing its target at random. These bird-hunting tactics had not been described before in the literature on the two species.

Another new hunting method of the Gyrfalcon – not described in the literature known to me – proved to be commonplace in a rural region of central Alberta. There, wintering Gyrfalcons hunted feeding Mallards by stealth approaches, but they also launched long-range climbing attacks on flocks approaching high in the sky and still >1 km away (Chapter 14). This active hunting method of the Gyrfalcon resembled the most spectacular duck-hunting flights I have ever seen of the Peregrine (Dekker 1999:129–131).

The safety aspect of flock formation. In the above described observations, it was very evident that the city pigeons, by massively flushing from the buildings or ground, actually made it easier for the falcons to affect a capture, whereas free-flying single pigeons pursued by Gyrfalcons escaped or were let go, except in one case in which a pigeon was chased and grabbed directly from behind. This pigeon probably had not seen the falcon coming (Chapter 13).

This suggests that the tendency of prey species to take to the air at the sight of a falcon and to form dense flocks – which is generally seen as an anti-predator strategy – not only draws the attention of the predator to the birds, but the birds seem to invite attention. In fact, flocks of shorebirds actually approach and track the falcon's movements (Buchanan et al. 1988). By doing so, they compromise the safety of the flock while individual birds find shelter inside the flock. This applies to Merlins and Peregrines attacking Dunlins at Boundary Bay, as well as to Prairie Falcons and Gyrfalcons hunting city pigeons.

The flocking habits of feeding shorebirds in regards to individual spacing and predation risk have been studied by several researchers (Cresswell 1994; Quinn and Cresswell 2004; Van den Hout et al. 2008). But it is an open question which class of prey – either the adults or the juveniles, or perhaps only the strongest flyers among them – find their way into the centre of flying flocks. Conversely, it remains to be determined which category of individuals is relegated to the less safe outside. As shown in this thesis, Peregrines attacking dense flocks of Dunlins tend to concentrate on the edges or bottom of flocks (Chapter 9).

Do parent Peregrines teach their fledgling how to hunt? The question of whether or not juvenile Peregrines have to be taught hunting skills by their parents has been raised before and answered convincingly by the releases of captive-raised juveniles that are now thriving in the wild (Cade and Burnham 2003). Apparently, chasing and grabbing hold of other birds comes naturally to young falcons. Nevertheless, there was a gap in detailed knowledge about parent-fledgling interactions (Newton 1979). In that regard, the observations reported here have added substantially to this debate (Chapter 10). Furthermore, this study has also shown convincingly that Peregrines gain in expertise and success rate as they become older, another subject about which, to my knowledge, there was no prior information available in the raptor literature.

A final note on the pesticide era, and the food base as the ultimate determinant in raptor dynamics. A question that was particularly relevant at the start of this study was whether the Peregrine would survive the pesticide era in view of the alarming reports about eggshell thinning and breeding failures. My findings of 35 years ago that immatures made up the majority (62–72%) of spring migrants came as a surprise (Chapter 2). Evidently, these northern migrants were still hatching young at what appeared adequate levels, which was later confirmed by intensive research conducted by government biologists on the coast of the Arctic Ocean. Evidently, despite rather high pesticide levels, Canadian tundra-breeding falcons continued to reproduce with erratic ups and downs that were found to be weather dependent and correlated with cyclic prey abundance (Court et al. 1990; Johnstone 1998).

The food resource also proved to be key to the decline of the Peregrine population that used to nest along Alberta's Red Deer River. As reported in this study, during the critical 1960s, when the local Peregrines became extirpated, humans had been robbing the young at all known traditional breeding sites. However, superimposed on these losses and the indirect role of agricultural poisons, I wondered about a possible third negative factor: habitat deterioration caused by wetland drainage and modern agricultural practices, which must have resulted in a declining prey base for the Peregrine (Dekker 1967). This problem became a most likely scenario by the late 1990s, after provincial and federal government agencies had organized a mass release of captive-raised falcons in the belief, and based on their own research, that the pesticide threat had lessened (Holroyd and Banasch 1990). My finding that the local breeding population collapsed again after an initial success, indeed makes it likely that environmental factors and a declining prey base were the underlying causes of the Peregrine's disappearance as a breeding bird along Alberta's prairie rivers (Chapter 12).

Concluding remarks. In addition to reporting on hunting habits and success rates, the publications that form the chapters in this dissertation have added new knowledge regarding several formerly misunderstood aspects of the biology and behaviour of the Peregrine, such as its migrations. For instance, the belief that falcons commonly travelled in flapping flight low over the ground (Brown and Amadon 1968) was contradicted by my observations that they often soared at altitudes exceeding 1 km (Chapter 2). High-altitude soaring by migrating Peregrines was later confirmed by the application of radio-telemetry (Cochran 1985). I also found that Peregrines soaring at altitudes of >1 km proved to be hunting and not just travelling. These falcons launched their surprise attacks by a near vertical stoop from a great height (Chapter 3).

Quite apart from amassing what is perhaps the world's largest sample of hunts and kills by Peregrines (Figure 16.1), the comparative value of the studies described in this dissertation might lie in the fact that they are, unlike many other studies, not limited to one location or brief time period. The wide variance found in hunting success rates and capture methods between the different periods and sites supports the view that the value of naturalistic field studies of wildlife become all the more ecologically meaningful the longer they are continued, (Newton 1979; Schmidly 2005), in particular if they are made by the same observer using similar methods throughout.

Table 16.1. Progression of total hunts and kills with comparative success rates. The data points are based on the time periods detailed in the major publications. The rising rate reflects a change in study sites when I began watching a breeding pair of Peregrines at Wabamun Lake, in addition to periodic observations of wintering and migrating falcons on the Pacific coast and at Beaverhills Lake.

Year	# of hunts	# of kills	Success rate
1980	674	52	7.7%
1987	1104	81	7.3
1994	1364	133	9.6
1996	1437	149	10.4
2003	2089	243	11.6
2004	2475	360	14.5
2008	3925	460	11.7

Table 16.2. Birds observed captured by wild falcons in Canada, 1965–2008.

	Peregrine	Merlin	Gyrfalcon	Prairie Falcon
Shorebirds	200	28	0	1
Gulls	88	0	0	0
Ducks	87	0	16	0
Passerines	41	16	0	1
Other or Unknown	24	0	0	0
Seabirds	16	0	0	0
Rock Doves	4	0	15	27
Total kills	460	44	31	29
Success rates	11.7	12.4	14.7	26.9

ACKNOWLEDGEMENTS

Although I have been a self-starter in all of my research projects, a number of individuals have helped me along the way over the course of 48 years. Very soon after arriving in my adoptive country, I joined the local bird club. Its general membership included university professors and government wildlife biologists, who introduced me to the ornithological literature of which I had been totally ignorant. For instance, the late Dr. Otto Hohn, upon learning of my interest in Peregrines, did me an important favour by supplying a copy of the classical 1950 paper on raptor foraging by the Swedish ornithologist Gustaf Rudebeck. It opened my eyes as to what could be done with my own growing list of field observations. Professor David Boag critically read my first manuscripts on Peregrines (published in 1979 and 1980). His comments taught me much. So did the subsequent extended process of editorial and referee input, which was of great help to improve my Dutch-accented English.

Editors and scientists who over the years reviewed my papers and made helpful comments were: James Bednarz, David Bird, Joseph Buchanan, Tom Cade, Francis Cook, Gordon Court, Cheryl Dykstra, Anthony Erskine, Stewart Houston, G. Malar, Wayne Nelson, Ian Newton, K.G. Poole, R. Ritchie, Ted Swem, Karin Uilhoorn, F. P. Ward, R. B. Weeden, Chip Weseloh, Clayton Whyte, and William Wishart.

Relevant literature was provided by Gerard Beyersbergen, Robert Butler, Rob Corrigan, David Henry, Otto Hohn, Wayne Nelson, Martin McNicholl, Jim Wolford, and Ronald Ydenberg. Long-time raptor-loving friend Gordon Court did some of the statistical tests for these papers. Other tests were done by my brother Marius, who is a genius in mathematics and did several of the graphs. Other graphs were drafted by my son Richard, who also set up the tables and continues to provide expert assistance with computer problems. Ingrid Luters of Hancock House Publishers skilfully scanned and inserted the black and white photos and some of the line drawings, as well as the colour images for the insert.

Comfortable accommodations during fieldwork in British Columbia were courtesy of David Hancock, Thea Leach, Edo Nyland, and the Schweers family. For access on private lands in Alberta I particularly want to thank Ron Goeglein (Beaverhills Lake), Lyall Russell, Nick & Louise Mullins (Red Deer River), and Ralph Leriger (TransAlta properties at Wabamun). Bob Elner of CWS Delta arranged dike access permits at Boundary Bay.

Quite apart from my dear wife Irma, who was and is in a class all by herself in providing support and encouragement, in the field as well as at home, I like to mention the following friends and acquaintances who occasionally acted as a second set of eyes during field observations. At the risk of forgetting one or two, they are in alphabetical order: Ludo Bogaert, Gordon Court, Marius Dekker, Cornelius Dekker, Albert Ferwerda, Brian Genereux, Kees Kunst, Jim Lange, Gerald Romanchuk, Ron Slagter, Rick Swanston, Miechel Tabak, and Robert Taylor.

Intermittent grants for partial reimbursement of travel expenses were obtained from the Alberta Fish and Wildlife Division, the Alberta Recreation, Parks & Wildlife Foundation, the Alberta Conservation Association, the Centre for Wildlife Ecology at Simon Fraser University, the Pacific Wildlife Research Centre of the Canadian Wildlife Service at Delta, and the Frisian nature conservation organisation *It Fryske Gea*.

In particular, I want to express my gratitude to Professors Ronald Ydenberg and Herbert Prins, who acted as my promoters in obtaining a degree of Doctor in Production Ecology and Resource Conservation from Wageningen University. They also taught me that writing a dissertation is quite different from writing a book for the general public, and they made numerous detailed and helpful comments on the four progressive drafts of the manuscript. As well, they guided me through the confounding process of compiling the *stellingen*. Thanks are also due to secretary Patricia Meijer for arranging an initial set of printed copies of the thesis. Mentioned last, but in fact worthy of first place, I am indebted to the Academic Board of Wageningen University, who exercised its right to admit me to the promotional process even though I lacked the conventional prerequisites in scholarly education.

The American Bald Eagle habitually klepto-parasitizes smaller raptors, including all four falcon species that are the subject of this dissertation. (Photo: David Hancock).

LITERATURE CITED

Anderson, C. M. and P. M. DeBruyn. 1979. Behavior and ecology of Peregrine Falcons wintering in the Skagit Flats. Report for the Washington Department of Game. Olympia, WA, USA.

Anderson, M. and R. A. Norberg. 1981. Evolution of reversed sexual dimorphism and role partitioning among predatory birds, with a size scaling of flight performance. Biological Journal of the Linnean Society 15:105–130.

Anderson, S. H. and J. R. Squires. 1997. The Prairie Falcon. University of Texas Press, Austin, TX, USA..

Baker, J. A. 1967. The Peregrine. Collins and Co. London, United Kingdom.

Beauvais, G., J. H. Enderson, and A. J. Magro. 1992. Home range, habitat use, and behavior of Prairie Falcons wintering in east-central Colorado. Journal of Raptor Research 26:13–18.

Beebe, F. L. 1960. The marine Peregrines of the northwest Pacific coast. Condor 62:145–189.

Beebe, F. L. 1974. Field studies of the falconiformes of British Columbia. British Columbia Provincial Museum. Occasional Paper Series No. 77.

Beebe, F. L. and H. M. Webster. 1964. North American Falconry and Hunting Hawsk. Private Publication. Denver, CO, USA.

Beemster, N. 1993. Gepaard overwinterende Slechtvalken in het Lauwersmeer gebied. De Grauwe Gors 21:52–55.

Bednarz, J. C. 1988. Cooperative hunting in Harris's Hawks (*Parabuteo unicinctus*). Science 239:1525–1527.

Bednekoff, P. A. 2007. Foraging in the face of danger. *Pages* 307-329 *in* Foraging: Behavior and Ecology. D. W. Stephens, J. S. Brown, and R. C. Ydenberg. *Eds*. The University of Chicago Press, IL, USA.

Bent, A. C. 1938. Life Histories of North American Birds of Prey. Parts 1 and 2. United States National Museum. Bulletin 170. Dover reprint 1961.

Bentson, S. A. 1971. Hunting methods and choice of prey of Gyrfalcons (*Falco rusticolus*) at Myvatn in Iceland. Ibis 113:468–476.

Berry, R. B. 1971. Peregrine Falcon population survey. Assateague Island, Maryland, USA. Fall 1969. Raptor Research News 5:31–43.

Bijlsma, R. G. 1990. Predation by large falcons on wintering waders on the Banque d'Arguin, Mauritania. Ardea 78:75–82.

Bird, D. M. and Y. Aubrey. 1982. Reproductive and hunting behaviour of Peregrine Falcons in southern Quebec. Canadian Field-Naturalist 96:167–171.

Bos, D., J. Feddema, en Y. van der Heide. 2007. De broedvogels van Noard-Fryslan Butendyks in 2006. Twirre 18:62–66.

Bradley, D. M. and L. W. Oliphant. 1991. The diet of Peregrine Falcons at Rankin Inlet, Northwest Territories: an unusually high proportion of mammalian prey. Condor 93:193–197.

Brandenburg, E. en I. Riemersma. 2002. Winterterritoria van de Slechtvalk in de Greidhoeke. Twirre 13:95–97.

Brennan, L. A., J. B. Buchanan, S. G. Herman, and T. M. Johnson. 1985. Interhabitat movements of wintering Dunlins in western Washington. Murrelet 66:11–16.

Brown, L. and D. Amadon. 1968. Eagles, Hawks, and Falcons of the World. Hamlyn House, GB.

Brown, J. S. and B. P. Kotler. 2007. Foraging and the Ecology of Fear. *Pages* 437 480 *in* Foraging, Behaviour an Ecology. D. W. Stephens, J. S. Brown, and R. C. Ydenberg. *Eds*. The University of Chicago Press, Chicago and London.

Buchanan, J. B. 1988. The effect of kleptoparasitic pressure on hunting behavior and performance of host Merlins. Journal of Raptor Research 22:63–64.

Buchanan, J. B. 1996. A comparison of behavior and success rates of Merlins and Peregrine Falcons when hunting Dunlins in two coastal habitats. Journal of Raptor Research 30:93–98.

Buchanan, J. B., S. G. Herman, and T. M. Johnson. 1986. Success rates of the Peregrine Falcon hunting Dunlin during winter. Raptor Research 20:130–131.

Buchanan, J. B., C. T. Schick, L. A. Brennan, and S. G. Herman. 1988. Merlin predation on wintering Dunlins: hunting success and Dunlin escape tactics. Wilson Bulletin 100:108-118.

Burns, J. G. and R. C. Ydenberg. 2002. The effects of wing-loading and gender on the escape flights of Least Sandpipers (*Calidris minutilla*) and Western Sandpipers (*Calidris mauri*). Behavioral Ecology and Sociobiology 52: 128–136.

Butler, R. W. and R. W. Campbell. 1987. The Birds of the Fraser River Delta: Populations, Ecology, and International Significance. Canadian Wildlife Service Occasional Paper Number 65. Ottawa, ON.

Butler, R. W. 1994. Distribution and abundance of Western Sandpipers, Dunlins, and Black-bellied Plovers in the Fraser River Estuary. *Pages* 18–23 *in* R. W. Butler and K. Vermeer. *Eds*. The abundance and distribution of estuarine birds in the Straight of Georgia, British Columbia. CWS Occasional Paper Number 83. Ottawa, ON.

Cade, T. J. 1960. Ecology of the Peregrine and Gyrfalcon populations in Alaska. University of California Publications in Zoology 63:151–290.

Cade, T. J. and R. Fyfe. 1970. The North American Peregrine Survey. Canadian Field-Naturalist 84:231–245.

Cade, T. J. 1982. The Falcons of the World. Cornell University Press, Ithaca NY, USA.

Cade, T. J., J. H. Enderson, C. G. Thelander, and C. M. White. 1988. *Eds.* Peregrine Falcon populations: their management and recovery. Proceedings of the 1985 International Peregrine Conference, Sacramento. Braun-Brumfeld, San Francisco, CA, USA.

Cade, T. J., P. Koskimies, and O. K. Nielsen. 1998. Gyrfalcon (*Falco rusticolus*). Birds of the Western Palearctic 2(1). Oxford University Press, GB.

Cade, T. J. 2003. Life history traits of the Peregrine in relation to recovery. *Pages* 2–11 *in* T. J. Cade and W. Burnham. *Eds*. Return of the Peregrine. The Peregrine Fund, Boise, ID, USA.

Cade, T. J. and W. Burnham. 2003. Return of the Peregrine. The Peregrine Fund, Boise, ID, USA.

Clum, N. J, and T. J. Cade. 1994. Gyrfalcon (*Falco rusticolus*). The Birds of North America, Number 114. A. Poole and F. Gill. *Eds*. The Academy of Natural Sciences, Washington, DC, USA.

Clunie, F. 1976. A Fiji Peregrine in an urban environment. Notornis 23:8–28.

Cochran, W. W. 1985. Ocean migration of Peregrine Falcons: Is the adult male pelagic? *Pages* 223–237 *in:* Proceedings of Hawk Migration Conference IV. M. Harwood. *Ed*. Hawk Migration Association of North America.

Corrigan, R. 2000. Survey of the Peregrine Falcon in Alberta. Species at risk report, No. 2. Alberta Sustainable Resource Development, Fish and Wildlife Division. Edmonton, AB, Canada.

Court, G. S., D. M. Bradley, C. C. Gates, and D. A. Boag. 1988a. The Population Biology of the Peregrine Falcon in the Keewatin District of the Northwest Territories, Canada. *Pages* 729–739 *in* T. J. Cade, J. H. Enderson, C. G. Thelander, and C. M. White. *Eds*. Peregrine Falcon Populations: their Management and Recovery. The Peregrine Fund, Boise, ID, USA.

Court, G. 1999. Edmonton's year of the gyr. Edmonton Naturalist 27:29–30.

Court, G. S., D. M. Bradley, C. C. Gates, and D. A. Boag. 1988b. Turnover and recruitment in a tundra population of Peregrine Falcons. Ibis 131:487–496.

Court, G. S., C. C. Gates, D. A. Boag, J. D. MacNeil, D. M. Bradley, A. C. Fesser, J. R. Patterson, G. B. Stenhouse, and L. W. Oliphant. 1990. A toxicological assessment of Peregrine Falcons, *Falco peregrinus tundrius*, breeding in the Keewatin District of the Northwest Territories, Canada. Canadian Field-Naturalist 104:255–272.

Court, G. S. 1993. A toxicological assessment of the American Peregrine Falcon (*Falco peregrinus anatum*) breeding in Alberta, Canada – 1968 to 1992. Alberta Fish and Wildlife Service, Occasional Paper No. 10.

Craighead, J. J. and F. C. Craighead. 1956. Hawks, Owls, and Wildlife. Dover Publications, New York, NY, USA.

Creel, S., D. Christianson, S. Liley, and J. A. Winnie. 2007. Predation risk affects reproductive physiology and demographics of elk. Science 315: 960.

Cresswell, W. 1994. Flocking is an effective anti-predation strategy in Redshanks. Animal Behaviour 47:433–442.

Cresswell, W. 1996. Surprise as a winter hunting strategy in Sparrowhawks, Peregrine Falcons, and Merlins. Ibis 138:684-692.

Cresswell, W. and D. P. Whitfield. The effects of raptor predation on wintering wader populations at the Tyninghame estuary, southeast Scotland. Ibis 136:223–232.

Cresswell, W. 2008. Review: Non-lethal effects of predation in birds. Ibis 150:3–17.

Dekker, D. 1967. Disappearance of the Peregrine Falcon as a breeding bird of a river valley in central Alberta. Blue Jay 25:175–176.

Dekker, D. 1977. Field identification of Peregrines, Prairie Falcons, and Gyrs in southern and central Alberta. Alberta Naturalist 7:1–5.

Dekker, D. 1979. Characteristics of Peregrine Falcons migrating through central Alberta, 1969-1978. Canadian Field-Naturalist 93:296–302.

Dekker, D. 1980. Hunting success rates, foraging habits, and prey selection of Peregrine Falcons migrating through central Alberta. Canadian Field-Naturalist 94:371–382.

Dekker, D. 1982. Occurrence and foraging habits of Prairie Falcons (*Falco mexicanus*) at Beaverhills Lake, Alberta. Canadian Field-Naturalist 96:477–478.

Dekker, D. 1983. Gyrfalcon sightings at Beaverhills Lake and Edmonton, Alberta, 1964–1983. Alberta Naturalist 13:103.

Dekker, D. 1984a. Prairie Falcon sightings in the Rocky Mountains of Alberta. Alberta Naturalist 14:48–49.

Dekker, D. 1984b. Spring and fall migrations of Peregrine Falcons in central Alberta, 1979–1983, with comparisons to 1969–1978. Journal of Raptor Research 18:92–97.

Dekker, D. 1984c. Migrations and foraging habits of Bald Eagles in east-central Alberta. 1964–1983. Blue Jay 42:199–205.

Dekker, D. 1985a. In Praise of the Bald Eagle. *Pages* 148–155 *in* Wild Hunters. CWD Publications, Edmonton, AB, CD.

Dekker, D. 1985b. Hunting behaviour of Golden Eagles migrating in south-western Alberta. Canadian Field-Naturalist 99:383–385.

Dekker, D. 1987. Peregrine Falcon predation on ducks in Alberta and British Columbia. Journal of Wildlife Management 51:156–159.

Dekker, D. 1988. Peregrine Falcon and Merlin predation on small shorebirds and passerines in Alberta. Canadian Journal of Zoology 66:925–928.

Dekker, D. 1991, 1998. Prairie Water – Wildlife at Beaverhills Lake, Alberta. University of Alberta Press, Edmonton, AB, CA.

Dekker, D. 1993. Valley of the Falcons. Alberta Naturalist 23:53–56.

Dekker, D. 1995. Prey capture by Peregrine Falcons wintering on southern Vancouver Island, British Columbia. Journal of Raptor Research 29:26–29.

Dekker, D. 1998a. Over-ocean flocking of Dunlins (*Calidris alpina*) and the effect of raptor predation at Boundary Bay, British Columbia. Canadian Field-Naturalist 112:694–697.

Dekker, D. 1998b. Wolven in de Wildernis. Uitgeverij Jan van Arkel, Utrecht, The Netherlands.

Dekker, D. 1999. Bolt from the Blue – Wild Peregrines on the Hunt. Hancock House Publishers, Surrey, BC, Blaine, WA, USA.

Dekker, D. 2001. Een Hollandse Woudloper in Canada. Uitgeverij Bosch & Keuning, Baarn, The Netherlands.

Dekker, D. 2002. The bighorn's dilemma. *Pages* 156–163 *in* Wildlife Adventures in the Canadian West. Rocky Mountain Books, Calgary, AB, CA.

Dekker, D. 2003. Peregrine Falcon predation on Dunlins and ducks and kleptoparasitic interference from Bald Eagles wintering in British Columbia. Journal of Raptor Research 37:91–97.

Dekker, D. 2008. Wildlife Investigations at Devona, Jasper National Park, for the winters of 2001–2006 with comparisons to 1981–2001. Prepared for Parks Canada, AB, CA.

Dekker, D. and L. Bogaert. 1997. Over-ocean hunting by Peregrine Falcons in British Columbia. Journal of Raptor Research 31:381–383.

Dekker, D. and J. Lange. 2001. Hunting methods and success rates of Gyrfalcons and Prairie Falcons preying on feral pigeons in Edmonton, Alberta. Canadian Field-Naturalist 155:395–401.

Dekker, D., and G. Court. 2003. Gyrfalcon predation on Mallards and the interaction of Bald Eagles wintering in central Alberta. Journal of Raptor Research 37:161–163.

Dekker, D. and R. Ydenberg. 2004. Raptor predation on wintering Dunlins in relation to the tidal cycle. Condor 106:415–419.

Dekker, D. and R. Taylor. 2005. A change in foraging success and cooperative hunting by a breeding pair of Peregrine Falcons and their fledglings. Journal of Raptor Research 39:386–395.

Dementiev, G. P. 1960. Der Gerfalke (*Falco rusticolus* L.). Die Neue Brehm Bucherei. Heft 264. Ziemsen Verlag. Wittenberg, Lutherstadt, Germany.

Dietz, M. W., T. Piersma, A. Hedenstrom, M. Brugge. 2007. Intraspecific variation in avian pectoral muscle mass: constraints on maintaining manoeuvrability. Functional Ecology 21: 317–326.

Dijkema, K. S., A. Nocolai, J. Frankes, H. Jongerius, en N. Nauta. 2001. Van landaanwinning naar kwelderwerken. Rijkswaterstaat Directie Noord Nederland, Leeuwarden. Alterra Research Institute voor de Groene Ruimte, Texel, The Netherlands.

Dobler, F. C. 1960. Wintering Gyrfalcon habitat utilization in Washington. *Pages* 61–70 *in* Raptors in the Modern World. B. Meyburg and R. D. Chancellor. *Eds*. World Working Group of Birds of Prey. Berlin, London, Paris.

Dunning, J. B. 1984. Body weights of 686 species of North American birds. Western Bird Banding Association. Monograph No. 1.

Ellis, D. H., J. C. Bednarz, D. G. Smith, and S. P. Flemming. 1993. Social foraging classes in raptorial birds. BioScience 43:14–20.

Enderson, J. H. 1964. A study of the Prairie Falcon in the central Rocky Mountain region. Auk 81:332–352.

Enderson, J. H. 1965. A breeding and migration survey of the Peregrine Falcon. Wilson Bulletin 77:327–339.

Enderson, J. H. 1969. Population trends among Peregrine Falcons in the Rocky Mountain region. *Pages* 73–79 *in* J. J. Hickey. *Ed.* Peregrine Falcon Populations: Their Biology and Decline. University of Wisconsin Press, Madison WI, USA.

Enderson, J. H. 1969. Peregrine and Prairie Falcon life tables based on band recovery data. *Pages* 505-509. *In* Peregrine Populations – Their Biology and Decline. J.J. Hickey. *Ed.* The University of Wisconsin Press, WI, USA.

Enderson, J. H. 2003. Early Peregrine expansion and management in the Rocky Mountain West. *Pages* 57–71 *in* T. J. Cade and W. Burnham. *Eds*. Return of the Peregrine. The Peregrine Fund, Boise, ID, USA.

Enderson, J. H. and G. R. Craig. 1997. Wide-ranging by nesting Peregrine Falcons determined by radio-telemetry. Journal of Raptor Research 31:333–338.

Engelmoer, M. 2001. Integrale tellingen van wadvogels langs de Friese Waddenkust tussen augustus 1994 en mei 2001 – Een tussenrapportage. Wadvogelwerkgroep, FFF.

Entermoser, A. 1961. Schlagen Beizvalken bevorzugt kranke kraehen? Vogelwelt 82:101-104.

Evans-Ogden, L. J., K. A. Hobson, D. B. Lank, and S. Bittman. 2005. Stable isotope analysis reveals that agricultural habitat provides an important dietary component for non-breeding Dunlins. Avian Conservation and Ecology.Online (URI:http://www.ace-eco.org/vol1/iss1/art3/Farr,A.1981) Foraging strategies of Dunlin in the intertidal areas between Boundary Bay and Roberts Bank. Unp. Ms. CWS, Delta, BC, CA.

Fischer, W. 1967. Der Wanderfalk. Die Neue Brehm Bucherei. Ziemsen Verlag. Germany.

Flater, D. (Online): Xtide:harmonic tide clock and tide predictor. http://www.flaterco.com/xtide (1 December 2003).

Frank, S. 1994. City Peregrines. A Ten-Year Saga of New York City Falcons. Hancock House Publishers, Surrey BC, Blaine WA.

Franklin, K. 1999. Vertical flight. NAFA Journal: 68–72. The American Falconers' Association.

Fyfe, R. W. 1976. Rationale and success of the Canadian Wildlife Service Peregrine breeding project. Canadian Field-Naturalist 90:309–319.

Fyfe, R. W., S. A. Temple, and T. J. Cade. 1976. The North American Peregrine Falcon survey. Canadian Field-Naturalist 90:228–273.

Godfrey, W. E. 1966. The Birds of Canada. National Museum of Canada. Bulletin 203. Ottawa, ON, CA.

Garber, C. S., B. D. Mutch, and S. Platt. 1993. Observations on wintering Gyrfalcons hunting Sage Grouse in Wyoming and Montana. Journal of Raptor Research 27:171–172.

Gosler, A. G., J. J. D. Greenwood, and C. Perrins. 1995. Predation risk and the cost of being fat. Nature 377:621–623.

Haugh, J. R. and T. J. Cade. 1966. The spring hawk migration around the southeastern shore of Lake Ontario. Wilson Bulletin 78:88–110.

Hector, D. P. 1986. Cooperative hunting and its relationship to foraging success and prey size in an avian predator. Ethology 73:247–257.

Herbert, R. A. and K. G. S. Herbert. 1965. Behavior of Peregrine Falcons in the New York City region. Auk 82:62–94.

Hicklin, P. W. 1987. The migration of shorebirds in the Bay of Fundy. Wilson Bulletin 99:540–570.

Hickey, J. J. (*Ed.*). 1969. Peregrine Falcon Populations – Their Biology and Decline. University of Wisconsin Press, WI, USA.

Holroyd, G. L. 2003. The return of the *anatum* Peregrine to southern Canada, 1985–2000. *Pages* 117–125 *in* T. J. Cade and W. Burnham. *Eds*. Return of the Peregrine Falcon. The Peregrine Fund, Boise, ID, USA.

Holroyd, G. L. and U. Banasch. 1990. The reintroduction of the Peregrine Falcon into southern Canada. Canadian Field-Naturalist 104:203–208.

Hotker, H. 2000. When do Dunlins spend high tide in flight? Waterbirds 23:482–485.

Holthuijzen, A. M. A. 1990. Prey delivery, caching, and retrieval rates in nesting Prairie Falcons. Condor 92:475–484.

Hosper, U. G. en J. de Vlas *(Redacteurs)*. 1994. Noord-Friesland Buitendijks. Beschrijving en toekomstvisie. Rapport Werkgroep NFB. It Fryske Gea, Otterterp, The Netherlands.

Hunt, W. G., R. R. Rogers, and D. J. Slowe. 1975. Migratory and foraging behavior of Peregrine Falcons on the Texas coast. Canadian Field-Naturalist 89:111–123.

Hustings, F. en E. van Winden. 1998a. Slechtvalken terug uit een diep dal. SOVON-Nieuws 11(4):14–16.

Hustings, F. en E. van Winden. 1998b. Slechtvalken buiten de broedtijd in Nederland. Slechtvalk Nieuwsbrief 4(2):2–6.

James, P. C. and A. R. Smith. 1987. Food habits of urban-nesting Merlins in Edmonton and Fort Saskatchewan, Alberta. Canadian Field-Naturalist 101:592–594.

Jenkins, A. R. 2000. Hunting mode and success of African Peregrine Falcons: Does nesting habitat quality affect foraging efficiency? Ibis 142:235–246.

Jenning, W. 1972. Iaktagelser rorande en overvintrande jaktfalk (*Falco rusticolus*). Var Fagelvarld 32:1–8.

Johnstone, R. M. 1998. Aspects of the population biology of Tundra Peregrine Falcons (*Falco peregrinus tundrius*). Ph. D. dissertation. University of Saskatchewan, Saskatoon, CN.

Kaufman, M. J., J. L. Pollock, and B. Walton. 2004. Spatial structure, dispersal, and management of a recovering raptor population. American Naturalist 164:582–597.

Kiff, L. F., R. G. Bijlsma, L. L. Severinghaus, and J. Shergalin. 2008. The Raptor Literature. *Pages* 11–33 *in* Raptor Research and Management Techniques. *Eds*. D. M. Bird and K. L. Bildstein. Hancock House Publishers, Surrey, B.C. Blaine, WA.

Klaassen, R. 1996. Slechtvalken in de uiterwaarden. Slechtvalk Nieuwsbrief 2(1):4–5.

Koks, B. 1998. Slechtvalken in de Nederlandse Waddenzee. Slechtvalk Nieuwsbrief 4(2):13–15.

Kus, B. E., P. Ashman, G. W. Page, and L. E. Stenzel. 1984. Age-related mortality in a wintering population of Dunlin. Auk 101:69–73.

Lange, J. 1985. Merlin kills pigeon. Alberta Naturalist 15:134.

Lange, J. 1995. Observing Gyrfalcons from the living room sofa. Edmonton Naturalist 23(2):12–13.

Lank, D. B., R. B. Butler, J. Ireland, and R. C. Ydenberg. 2003. Effects of predation danger on migratory strategies of sandpipers. Oikos 103:303–319.

Lima, S. L. 1987. Vigilance while feeding and its relation to the risk of predation. Journal of Theoretical Biology 124:303–316.

Lima, S. L. 1988. Vigilance and diet selection: a simple example in the Dark-eyed Junco. Canadian Journal of Zoology 66:593–596.

Lima, S. L., and L. M. Dill. 1990. Behavioral decisions made under the risk of predation – A review and prospectus. Canadian Journal of Zoology 68:619–640.

Marcus, P. 2005. Overwinterende Slechtvalken in Amsterdam en omgeving. Slechtvalk Nieuwsbrief 11(1):8–15.

Mawhinney, K. P., P. W. Hicklin, and J. S. Boates. 1993. A re-evaluation of the numbers of migrant Semipalmated Sandpipers, *Calidris pusilla*, in the Bay of Fundy during fall migration. Canadian Field-Naturalist 107:19–23.

McEwan, E. H., and P. M. Whitehead. 1984. Seasonal changes in body weight and composition in Dunlin (*Calidris alpina*). Canadian Journal of Zoology 62:154–156.

Mech, D. L. 1966. The Wolves of Isle Royale. United States Fauna Series 7.

Mech, L. D. and R. O. Peterson. 2003. Wolf-prey relations. *Pages* 131–157 *in* Wolves, behaviour, ecology, and conservation. L. D. Mech and L. Boitani. *Eds*. The University of Chicago Press, IL. USA.

Monneret, R. J. 1973. Techniques du chasse du Faucon Pelerin (*Falco peregrinus*) dans une region de moyenne montagne. Alauda 41:403–412.

Morrison, R. I. G. 1977. Use of the Bay of Fundy by shorebirds. *Pages* 187–199 *in* Fundy Tidal Power and the Environment. *Ed*. G. A. Daborn. The Acadian University Institute. Wolfville NS, CN.

Mueller, H. C. and D. D. Berger. 1961. Weather and fall migration of hawks at Cedar Grove, Wisconsin. Wilson Bulletin 73:171–192.

Nelson, M. W. 1969. The status of the Peregrine Falcon in the northwest. *Pages* 61–72 *in* J. J. Hickey. *Ed*. Peregrine Falcon Populations: their Biology and Decline. University of Wisconsin Press, Madison, WI, USA.

Nelson. R. W. 1970. Some aspects of the breeding behaviour of Peregrine Falcons on Langara Island, British Columbia. M.Sc. thesis. University of Calgary, AB, CN.

Nelson, R. W. 1990. Status of the Peregrine Falcon, *Falco peregrinus pealei*, on Langara Island, British Columbia, 1968–1989. Canadian Field-Naturalist 104:193–199.

Newton, L. 1979. Population Ecology of Raptors. Buteo Books, Vermillion, SD, USA.

Newton, I. 1998. Population Limitation in Birds. Academic Press, San Diego, CA, USA.

Newton, I. 2003. The contribution of Peregrine research and restoration to a better understanding of Peregrines and other raptors. *Pages* 339–347 *in* T. J. Cade and W. Burnham. *Eds*. Return of the Peregrine. The Peregrine Fund, Boise, ID, USA.

Oliphant, L. W. 1991. Hybridization between a Peregrine Falcon and a Prairie Falcon in the wild. Journal of Raptor Management 25:36–39.

Oosterhuis, R. en J. Kok. 2003. Voedselkeuze van de Slechtvalk op Griend. Twirre 14:105–108.

Ouweneel, G. L. 1995. Voorkomen en terreinkeus van de Slechtvalken in het noordelijke Delta gebied. Vogeljaar 43:193–198.

Page, G. 1974. Age, sex, molt, and migration of Dunlins at Bolinas Lagoon. Western Birds 5:1–12.

Page, G. and D. F. Whitacre. 1975. Raptor predation on wintering shorebirds. Condor 77:73–83.

Palleroni, A., C. T. Miller, M. Hauser, and P. Marler. 2005. Prey plumage adaptation against falcon attack. Nature 434:973–974.

Palmer, R. S. *Ed*. Handbook of North American Birds. Volumes 4 and 5 Diurnal Raptors. Yale University, New Haven, London.

Parker, A. 1979. Peregrines at a Welsh coastal eyrie. British Birds 72:104–114.

Peakall, D. B., D. G. Noble, J. E. Elliott, J. D. Somers, and G. Erickson. 1990. Environmental contaminants in Canadian Peregrine Falcons – A toxicological assessment. Canadian Field-Naturalist 104:244–254.

Piersma, T. 1994. Close to the edge: Energetic bottlenecks and the evolution of migratory pathways in Knots. Ph. D. thesis. Rijksuniversiteit Groningen, The Netherlands.

Piersma, T, A. Koolhaas, en J. Jukema. 2003. Seasonal body mass changes in Eurasian Golden Plovers staging in The Netherlands: decline in late autumn mass peak correlates with increase in raptor numbers. Ibis 145:565–571.

Platt, J. B. 1976. Gyrfalcon nest site selection and winter activity in the western Canadian arctic. Canadian Field-Naturalist 90:338–345.

Poole, K. G. and D. A. Boag. 1988. Ecology of Gyrfalcons in the central Canadian arctic: Diet and feeding behaviour. Canadian Journal of Zoology 66:334–344.

Porter, R. D. and C. M. White. 1973. The Peregrine Falcon in Utah, emphasizing ecology and competition with the Prairie Falcon. Brigham Young University, Science Bulletin. Biological Series 18:1–74.

Potapov, E. and R. Sale. 2005. The Gyrfalcon. T. & AD Poyser, UK, and Yale University Press, USA.

Powell, G. V. N. 1974. Experimental analysis of the social values of flocking by Starlings (*Sturnus vulgaris*) in relation to predation and foraging. Animal Behavior 22:501–505.

Prins, H. H. T. and G. Iason. 1989. Dangerous lions and nonchalant buffalo. Behaviour 108:262–296.

Ratcliffe, D. A. 1962. Breeding density in the Peregrine Falcon (*Falco peregrinus*) and Raven (*Corvus corax*). Ibis 104:13–39.

Ratcliffe, D. A. 1963. The status of the Peregrine in Great Britain. Bird Study 10:56–90.

Ratcliffe, D. A. 1967. Decrease in eggshell weight in certain birds of prey. Nature 215:208–210.

Ratcliffe, D. 1980. The Peregrine Falcon. Buteo Books, Vermillion, SD USA.

Ratcliffe, D. 1993. The Peregrine Falcon. Academic Press, San Diego, CA USA.

Roalkvam, R. 1985. How effective are hunting Peregrines? Journal of Raptor Research 19:27–29.

Rose, B. J. 1965. Notes on a Peregrine Falcon and Franklin's Gull encounter. Blue Jay 28:163.

Pomeroy, A. C. 2006. Trade-offs between food abundance and predation danger in spatial usage of a stopover site by western sandpipers). Oikos 112:629–637.

Pomeroy, A. C., R. W. Butler, and R. C. Ydenberg. 2006. Experimental evidence that migrants adjust usage at a stopover site to trade off food and danger. Behavioral Ecology 17:1041–1045.

Powell, G. V. N. 1974. Experimental analysis of the social value of flocking by Starlings (*Sturnus vulgaris*) in relation to predation and foraging. Animal Behaviour 22:501–505.

Quinn, J. L. and W. Cresswell. 2004. Predator hunting behaviour and prey vulnerability. Journal of Animal Ecology 73:143–154.

Rowell, P., G. L. Holroyd, and U. Banash. 2003. The 2000 Canadian Peregrine Falcon Survey. Journal of Raptor Research 37:98–116.

Rudebeck, G. 1950, 1951. The choice of prey and modes of hunting of predatory birds with special reference to their selective effect. Oikos 2:65–88; 3:201–231.

Selander, R. K. 1966. Sexual dimorphism and differential niche utilization in birds. Condor 68:113–151.

Semenchuk, G. P. 1992. The Atlas of Breeding Birds of Alberta. Federation of Alberta Naturalists, Edmonton, AB, CN.

Sherrod, S. 1983. Post-fledgling behavior of the Peregrine Falcon. The Peregrine Fund, Ithaca, NY, USA.

Sherrod, S. K. 1988. Behavioral differences in Peregrine Falcons. *Pages* 741–748 *in* T. J. Cade, J. H. Enderson, C. G. Thelander, and C. M. White. *Eds*. Peregrine Falcon Populations. The Peregrine Fund Inc., Boise, ID, USA.

Shor, W. 1970. Peregrine Falcon population dynamics deduced from band recovery data. Raptor Research 4:49–59.

Schmidly, D. J. 2005. What it means to be a naturalist and the future of natural history at American universities. Journal of Mammalogy 86:449–456.

Smit, H. 2000. Prooien van Slechtvalken op de Engelsmanplaat. Slechtvalk Nieuwsbrief 6(2):3–4.

Smith, D. W, D. R. Stahler, D. S. Guernsey, M. Metz, E. Albers, L. Williamson, N. Legere, E. Almberg, and R. McIntyre. 2007. Yellowstone Wolf Project, Annual Report 2007. National Park Service Publication YCR-2008-01. Yellowstone National Park, WY, USA.

Sokal, R. R. and F. J. Rohlf. 1969, 1981. Biometry: the principles and practice of statistics in biological research. Freeman and Company, San Francisco, CA, USA.

Steenhof, K. 1998. Prairie Falcon (*Falco mexicanus*) *in* The Birds of North America, No. 346. A. Poole and F. Gills. *Eds*. The American Ornithologists' Union, Washington, DC, USA.

Steenhof, K., M. N. Kochert, L. B. Carpenter, and R. N. Lehman. 1999. Long-term Prairie Falcon population changes in relation to prey abundance, weather, land use, and habitat conditions. Condor 101:28–41.

Stephens, D. W., J. S. Brown, and R. C. Ydenberg. 2007. Foraging, Behavior and Ecology. The University of Chicago Press, Chicago, IL, USA.

Teunissen, B. 1997. Slechtvalken in de Ackerdijkse Plassen. Slechtvalk Nieuwsbrief 3(1):4–5.

Thiollay, J. M. 1982. Les ressources alimentaires, facteur limitant la reproduction d'une population insulaire de Faucons Pelerins (*Falco peregrinus brookei*). Alauda 50:16–44.

Thiollay, J. M. 1988. Prey availabiliy limiting an island population of Peregrine Falcons in Tunesia. *Pages* 701–710 *in* T. J. Cade, J. H. Enderson, C. G. Thelander, and C. M. White. *Eds*. Peregrine Falcon populations: their management and recovery. The Peregrine Fund, Boise, ID, USA.

Tinbergen, N. 1951. The Study of Instinct. Oxford University Press, GB.

Tordoff, H. B. and P. T. Redig. 2003. Peregrines in the Midwest. *Pages* 137–187 *in* T. J. Cade and W. Burnham. *Eds*. Return of the Peregrine. The Peregrine Fund, Boise, ID, USA.

Treleaven, R. B. 1961. Notes on the Peregrine in Cornwall. British Birds 54:136–142.

Treleaven, R. B. 1980. High and low intensity hunting in raptors. Zeitschrift fur Tierpsychology 54:339–345.

Treleaven, R. B. 1998. In Pursuit of the Peregrine. Tiercel Publishing, Wheathampstead, Herts, GB.

Turner, B. C., G. S. Hochbaum, F. D, Caswell, and D. J. Newman. 1987. Agricultural impacts on wetland habitats on the Canadian prairies, 1981–1985. Transactions of the North American Wildlife and Natural Resources Conference 52:206–215.

Valone, T. J. and S. L. Lima. 1987. Carrying food items to cover for consumption: the behaviour of ten bird species under the risk of predation. Oecologia 71:286-294.

Van Dijk, J. 1996. Overwinterende Slechtvalken bij Zwolle. Slechtvalk Nieuwsbrief 2(1):10–11.

Van Dijk, J. 1997. Slechtvalken bij Zwolle; een overzicht van 1994 t/m mei 1997 (met een lijst van 666 prooien). Slechtvalk Nieuwsbrief 3(2):6–7.

Van Dijk, J. 2000. Zwolse Slechtvalken op middelbare leeftijd (met een lijst van 1,365 verzamelde prooien). Slechtvalk Nieuwsbrief 6(2):6–10.

Van Geneijgen, P. en J. van Dijk. 1997. De postduif in het slechtvalkmenu. Slechtvalk Nieuwsbrief 3(2):8–10.

Van Geneijgen, P. 2005. Broedresultaten van Slechtvalken in Nederland in 2005. Slechtvalk Nieuwsbrief 11:2–7.

Van Geneijgen, P. 2006. Broedresultaten van Slechtvalken in Nederland in 2006. Slechtvalk Nieuwsbrief 12:2–9.

Van den Hout, P., B. Spaans, and T. Piersma. 2008. Differential mortality of wintering shorebirds on the Banc d'Arguin, Mauritania, due to predation by large falcons. Ibis 150:219–230..

Vasina, W. G. and R. J. Straneck. 1984. Biological and ethological notes on *Falco peregrinus cassini* in central Argentina. Raptor Research 18:123–130.

Walton, B. J. 1978. Peregrine–Prairie Falcon interaction. Raptor Research 12:46–47.

Ward, F. P. and R. B. Berry. 1972. Autumn migrations of Peregrine Falcons on Assateague Island, 1970-1971. Journal of Wildlife Management 36:484–492.

Warnock, N. D. 1994. Biotic and abiotic factors affecting the distribution and abundance of a wintering population of Dunlin. Ph. D. dissertation. University of California, Berkely CA, USA.

Warnock, N. D., and R. E. Gill. 1996. Dunlin (*Calidris alpina*). *In* A Poole and F. Gill. *Eds*. The Birds of North America. No. 203. The Academy of Natural Sciences, Philadelphia, and the American Ornithologists' Union, Washington, D.C, USA..

White, C. M. 1968. Diagnosis and relationships of the North American tundra inhabiting Peregrine Falcons. Auk 85:179–191.

White, C. M. 1975. Studies on Peregrine Falcons in the Aleutian Islands. *Pages* 33–50 *in* J. R. Murphy, C. M. White, and B. E. Harrell. *Eds*. Population status of raptors. Raptor Research Report No. 3. Vermillion, SD, USA.

Wheeler, B. K. and W. S. Clark. 1995. North American Raptors. Academic Press Limited. London, GB, and San Diego, USA.

White, C. M. and R. B. Weeden. 1966. Hunting methods of Gyrfalcons and behavior of their prey. Condor 68:517–519.

White, C. M. and R. W. Nelson. 1991. Hunting range and strategies of tundra-breeding Peregrine and Gyrfalcons observed from a helicopter. Journal of Raptor Research 25:49–62.

White, C. M., D. J. Brimm, and J. H. Wetton. 2000. The Peregrine Falcon in Fiji and Vanuata. *Pages* 707–720 *in* R. D. Chancellor and B. U. Meyburg. *Eds*. Raptors at Risk. Hancock House Publishers, Blaine, WA, USA.

White, C. M., N. J. Clum, T. J. Cade, W. G. Hunt. 2002. Peregrine Falcon (*Falco peregrinus*). *In* A. Poole and E. Gill. *Eds*. The Birds of North America, No. 660. The Birds of North America Inc. Philadelphia, PA USA.

Whitfield, D. P. 1985. Raptor predation on wintering waders in southeast Scotland. Ibis 127:544–558.

Wokke, E. E. 2002. Overwinterende Slechtvalken in de Haarlemermeer, 2001–2002. Slechtvalk Nieuwsbrief 8(2):2–6.

Ydenberg, R. C. and H. H. T. Prins. 1984. Why do birds roost communally? *Pages* 123–239 *in* P. R. Evans, J. D. Goss-Custard, and W. G. Hale. *Eds*. Coastal waders and wildfowl in winter. Cambridge University Press, Cambridge, UK.

Ydenberg, R. C., R. W. Butler, D. B. Lank, C. G. Guglielmo, M. Lemon, and N. Wolf. 2002. Trade-offs, condition dependence and stop-over site selection by migrating sandpipers. Journal of Avian Biology 33:47–55.

Ydenberg, R. C., R. W. Butler, D. B. Lank, B. D. Smith, and J. Ireland. 2004. Western Sandpipers have altered migration tactics as Peregrine Falcon populations have recovered. Proceedings Royal Society, Series B 271: 1263–1269. London, GB.

Ydenberg, R. C., R. B. Butler, and D. B. Lank. 2007. Effects of predator landscapes on the evolutionary ecology of routing, timing and moult by long-distance migrants. Journal of Avian Biology 38:523–529.

Ydenberg, R., D. Dekker, G. Kaiser, P. Sheppard, L. Evans-Ogden, K. Richards, and D. Lank. 2009. Winter mass and high tide behaviour as danger management tactics by wintering Pacific Dunlins. In submission.

CURRICULUM VITAE

Theodorus Johannes (Dick) Dekker was born in Rotterdam on 19 October 1933. Just after the outbreak of the Second World War and before Rotterdam was bombed, he moved to Haarlem, where he received his secondary education at Het Triniteitslyceum. In 1950 he began a career in graphic design and publishing at Uitgeverij De Spaarnestad. Nature study had become a lifelong passion. At 17, he sold his first illustrated articles to youth magazines and eventually to a wide range of print media. His list of publications in the Dutch language, which continues to grow, now includes 222 titles and seven books. In 1959 he emigrated to Canada in search of unspoiled wilderness. After almost having lost his life in a canoe accident in Yukon, he returned to Haarlem in 1961, resuming his nature writing until 1964, when he again left for Canada, accompanied by his wife Irma. In Edmonton, Alberta, he established himself as a free-lance graphic designer until 1982. Since then he has devoted himself full-time to wildlife writing and field research, which includes long-range mammal surveys in Jasper National Park, and ongoing studies of the avian inventory and habitat succession at Beaverhills Lake, a Ramsar wetland and Hemispheric Shorebird Reserve. He has given talks about wolves and other wildlife at a number of national and international conferences and nature group meetings. His growing list of publications in the English language now sits at 180 titles, including 27 papers in refereed journals, nine books, and three scripts for television documentaries. Throughout, his main focus has been on the dynamics of predator and prey interactions, with particular emphasis on the Peregrine and other falcons.

In celebration of the return of the Peregrine.
Dedicated to those who made it happen.

During their spring migration through central Alberta, Peregrines commonly make a stopover on the open pastures around Beaverhills Lake. Arriving in late afternoon or early evening, they tend to seek a perch on prominent points and often select a fence post standing in water. In order to find out where these falcons roost for the night, I watched a number of them until it became too dark to see. To my surprise, at dusk some Peregrines took off to hunt ducks, which at that time of day leave the lake and head inland. Flying low over the dark fields, the falcons swooped upward at prey outlined against the still luminous sky. I suspect that night hunting also is a way of escaping the attention of scavengers intent on robbing falcons of their catch.

If attacked by a Peregrine, flying ducks habitually save themselves by plunging into the nearest water. The immature falcon in the painting at left, in hot pursuit of four Northern Pintails (*Anas acuta*), is not likely to catch one of them, because the ducks are already over water and descending. By contrast, as illustrated in the painting above, this falcon is coming in low from the lake and in a good position to overtake its prey over land, grabbing it in midair or after it drops down onto marshy ground. Migrating Peregrines consume only the neck and part of the breast of a duck and leave the remains to scavengers. Females are capable of warding off harriers and buzzards, but male Peregrines – one third smaller than the females – commonly lose their ducks.

Sandpipers select open feeding grounds that allow a wide field of view, and they crowd together according to the principle that there is safety in numbers and that two pairs of eyes see more than one pair. Falcons that hunt shorebirds commonly attack by stealth, using vegetation to stay out of sight until the very last moment when the prey flushes in alarm. At Alberta's Beaverhills Lake, Merlins used stealth methods in 72% of 354 observed attacks, and Peregrines in 74% of 647 attacks. Merlins hunting small shorebirds succeeded in 12.5% of attacks, and Peregrines in 8.8%, but the difference was not statistically significant. If surprise attacks failed, both falcons approached the shorebirds openly and might pursue a selected target high into the sky, making repeated swoops until the prey is either caught or escapes.

The painting at left shows a Merlin in a typical stealth attack on Least Sandpipers (*Calidris minutilla*). Above, an adult Peregrine about to take a Black-bellied Plover (*Pluvialis squatarola*) by surprise.

Along a 15-km stretch of Alberta's Red Deer River that I named the Valley of the Falcons, I found three or four breeding pairs each of Peregrines and Prairie Falcons in 1960. By the end of that decade, the former had become extirpated, while the latter continued to produce young. The causes of the Peregrine's demise were no doubt related to pesticide pollution and direct human interference, but I wondered whether there were other factors involved, namely interspecific competition for nest sites and habitat change resulting in a declining prey base for the Peregrine. An opportunity to test this hypothesis arrived in the 1990s, when Canadian government agencies released about 200 captive-raised Peregrines at hack-sites along the river. After an initial success, resulting in the return of the Peregrine as a local breeding bird, the species dwindled again to only one nest in 2005–2008. In an inverse relationship, the Prairie Falcons increased. The painting at left depicts a pair of prairies soaring over their traditional eyrie cliffs.

Above, a Golden Eagle in power glide along a foothills slope. A versatile hunter, it mainly captures ground squirrels and rabbits, but it is quite capable of seizing ducks in flight if they flush just ahead. It is also a habitual klepto-parasite, robbing smaller raptors of their catch, but this is a trait shared by all wild carnivores.

The American Bald Eagle is a grand example of the carnivore motto: *why exert yourself if you can eat for free*. However, the universal principle is energy efficiency. Capable of killing waterfowl up to the size of swans, the Bald Eagle is usually content to feed on fish or even carrion. It is also quick to take advantage of smaller raptors. On the Pacific coast of western Canada, the ubiquitous presence of Bald Eagles forces wintering Peregrines to stop hunting ducks, and to concentrate on sandpipers that can be carried off at the approach of klepto-parasites. In central Alberta, eagles keep watch at stretches of river – downstream from big cities – that stay unfrozen and where Mallards (*Anas platyrhynchos*) overwinter. Although they manage to catch a few ducks themselves, the eagles wait for Gyrfalcons to bring down Mallards that fly inland to feed on grain.

Paintings © Dick Dekker. First published in: Dekker. D. 1980. Naturalist Painter.
Western Producer Prairie Books. Saskatoon, SK Canada

www.ingramcontent.com/pod-product-compliance
Lightning Source LLC
Chambersburg PA
CBHW080539300426
44111CB00017B/2797